Coincidences, Chaos, and All That Math Jazz

Coincidences, Chaos, and All That Math Jazz

MAKING LIGHT OF WEIGHTY IDEAS

Edward B. Burger
& Michael Starbird

Original art by Alan Witschonke Illustration

W. W. NORTON & COMPANY   NEW YORK   LONDON

For information about permission to reproduce selections from this book, write to
Permissions, W. W. Norton & Company, Inc., 500 Fifth Avenue, New York, NY 10110

Manufacturing by Courier Westford
Book design by Soonyoung Kwon
Production manager: Julia Druskin

Library of Congress Cataloging-in-Publication Data

Burger, Edward B., 1963–
    Coincidences, chaos, and all that math jazz : making light of weighty ideas /
Edward B. Burger, Michael Starbird ; original illustrations by Alan Witschonke.—1st ed.
        p. cm.
    Includes bibliographical references and index.
    ISBN 0-393-05945-6 (hardcover)
1. Mathematics—Humor. I. Starbird, Michael P. II. Title.
    QA99.B87 2005
    510—dc22

                                                              2005011106

W. W. Norton & Company, Inc., 500 Fifth Avenue, New York, N.Y. 10110
www.wwnorton.com

W. W. Norton & Company Ltd., Castle House, 75/76 Wells Street, London W1T 3QT

1 2 3 4 5 6 7 8 9 0

CONTENTS

Opening Thoughts   *vii*

PART I **UNDERSTANDING UNCERTAINTY**
Coincidences, Chaos, and Confusion   I

1– **UNBRIDLED COINCIDENCES**
Likelihood, Lady Luck, and Lady Love   3

2– **CHAOS REIGNS**
Why We Can't Predict the Future   20

3– **DIGESTING LIFE'S DATA**
Statistical Surprises   42

PART II **EMBRACING FIGURES**
Sensing Secrecy, Magnificent Magnitudes,
and Nature's Numbers   63

4– **SECRETS HELD, SECRETS REVEALED**
Cryptography Decrypted   65

5– **SIZING UP NUMBERS**
How Many? How Big? How Quick?   78

6– **A SYNERGY BETWEEN NATURE AND NUMBER**
A Search for Pattern   100

PART III   **EXPLORING AESTHETICS**
Sexy Rectangles, Fiery Fractals,
and Contortions of Space   121

7– **FROM PRECISE BEAUTY TO PURE CHAOS**
Picturing Aesthetics Through the Lens of Mathematics 123

8– **ORIGAMI FOR THE ORIGAMICALLY CHALLENGED**
From Paper Folding to Computers and
Fiery Fractals   146

9– **A TWISTED TURN IN AN AMORPHOUS UNIVERSE**
An Exploration of an Elasticized World   166

PART IV   **TRANSCENDING REALITY**
The Fourth Dimension and Infinity   199

10– **THE UNIVERSE NEXT DOOR**
The Magic of the Fourth Dimension   201

11– **MOVING BEYOND THE CONFINES OF
OUR NUTSHELL**
A Journey Into Infinity   231

12– **IN SEARCH OF SOMETHING STILL LARGER**
A Journey Beyond Infinity   246

Closing Thoughts   267
Acknowledgments   268
Further Resources   269
Permissions   270
Index   271

We believe that all curious people can enjoy and understand great mathematical ideas without having to brush up on garden-variety school math or relive their painful algebra daze. The incredible ideas of mathematics can tickle the imagination and open the mind. We have tried to make *Math Jazz* a place where math-o-philes and math-o-phobes alike will enjoy a refreshing, lighthearted adventure. If the sight of an equation makes you ill, this is the mathematics book for you! All inquiring minds can understand and enjoy intriguing ideas such as coincidences and chaos and the fourth dimension. Seemingly inexplicable and inaccessible concepts such as infinity and public key cryptography are within everyone's grasp. Here we bring these lofty notions down to earth and try to make them comprehensible and even (dare we say it?) fun.

Often the way to understand the true essence of all this math jazz is to look closely at simple and familiar features of our everyday world. We "make light of weighty ideas" by drawing attention to ordinary aspects of everyday life that normally go unnoticed and exploring their consequences. What leads to deep mathematical ideas? Counting the spirals on the prickly façades of pineapples and pine cones or looking closely at the chaotic creases created by folding a piece of paper. Surprisingly quickly we can move from a silly observation to a profound mathematical insight. A little logical thinking goes a long, long way. After we see a pattern on a pine cone, just a few easy steps take us to the discovery of a number pattern that has an organic life of its own and expresses itself in paintings, architecture, and music. We love to find atoms of clarity that arrange themselves into intriguing conceptual shapes. At their core, higher mathematical ideas are neither inaccessible nor incomprehensible. Deep ideas often have quite simple origins.

In our experience of presenting this material, people respond, "I love this stuff. But where's the math?" What we present may not resemble math, because we avoid the cryptic equations, formulas, and graphs that many people have come to know and fear as mathematics. Indeed, those symbols are the memorable icons of an often-forbidding foreign language of mathematical jargon, but it's not the only language of mathematics and it does not reside at the center of the subject. The deepest and richest realms of mathematics are often devoid of the cryptic symbols that have baffled students through the generations. Ideas—intriguing, surprising, fascinating, and beautiful—are truly at the heart of mathematics.

Many people think mathematics is the mechanical pursuit of solving equations. In truth, mathematics is an artistic pursuit. To mathematicians, it's a world of truths that can be established through imaginative proofs woven with threads of subtle logic. Here we "make light of weighty ideas" by taking readers on a lively journey of imagination and abstraction where serious mathematical ideas are presented in sometimes irreverent ways. But no one should be fooled into believing that the lighthearted tone implies that we are not pursuing lofty goals. Within these pages is authentic mathematics, often of a rather advanced kind, but presented in a way that enlists the help of our (and your) everyday experiences. Our hope is that these puzzles, stories, and illustrations will stimulate great discussions and debates over dinner and cocktails.

The common threads among all the adventures in this book are the intriguing surprises that arise organically from four fundamental aspects of human experience: grappling with uncertainty, counting and quantification, visualizing our physical world, and transcending our everyday world. In fact, through our journey, we will discover what a *surprise* truly is—a moment in which our intuition runs counter to reality. Here we'll try to set our intuition straight.

We have included some of our own favorite themes that have been a source of intrigue and delight for our non-mathematical audiences and ourselves over the years. Uncertainty accompanies every step of our lives, and the notions of coincidences, chaos, and statistics all surprise us with counterintuitive, solid insights in the shaky world of random chance. Counting is a basic way to see our world with more precision. But there's nothing basic or mundane involved in skulking through the secret world of cryptography or grappling with numbers too large to count or finding numerical patterns on pineapples. Our visual world opens our eyes to beauty, detail, and form. The Golden Rectangle, the fiery Dragon Curve fractal, and a twisted world of unlimited elasticity all present us with visual intrigue. The book closes by transcending reality itself. We journey into the alien worlds of the fourth dimension and infinity and hope we make it home in one mental piece.

The simple but clear thinking that allows us to discover the secrets of mathematics can help us resolve the conundrums of real life as well. The creative mindset embodied by mathematics enables us to see our ordinary world in a deeper way. Thus a journey through abstract thought and unbridled imagination provides a beautiful and powerful beacon for our journey through life.

—Edward Burger and Michael Starbird
January 1, 2005

Coincidences,
Chaos, and
All That Math Jazz

PART I

UNDERSTANDING UNCERTAINTY

Coincidences, Chaos, and Confusion

We open with a wild ride through the world of randomness and chance—a world that combines sheer volatility with astonishing tameability. Our destinies are largely determined by the whims of random luck (what we often think of as just plain dumb luck). Here, in Part I, we look at three faces of chance: coincidence, which involves unexpected convergence; chaos, which involves unexpected divergence; and statistics, which is an attempt to measure the uncertain in a fair and balanced way. Each of these themes is full of surprises that challenge our intuitions as we ride the random roller coaster we know as life.

Coincidences are our constant companions in our everyday lives—lives that hang heavily on randomness and chance. As we'll see, if we were to randomly kidnap 35 people off the street, two events are remarkably likely to happen—we'd get arrested, and we'd experience a surprising coincidence, for two of those 35 plaintiffs would almost surely share the same birthday. We'll see why monkeys could eventually produce this exact book, if we just let them aimlessly bash on a keyboard longer than any publisher is willing to wait. Here we'll see why surprises aren't all that surprising and that we should expect the unexpected.

Chaos, on the other hand, which is also a fundamental part of our lives, represents unexpected divergence. Many people—especially those with children—view "chaos" as a complete loss of control. But here we will meet mathematical chaos, which is surprisingly orderly and understandable, and yet leads to total unpredictability. We'll see why even error-free computers will sometimes give us wildly wrong answers. Imperceptibly small differences at the outset can often snowball into dramatically different end results. Thus we will come to understand why the local TV news offers us a paltry five-day weather forecast and will never present the more informative thirty-day forecast. Butterflies gently flapping their wings will continue to confound even the mightiest of meteorologists. Chaos not only affects the weather but also injects itself in nearly every circumstance that requires us to apply current conditions to predict the future. Chaos reigns.

With our heads spinning in this world of coincidence and chaos, we nevertheless must make decisions and take steps into the minefield of our future. To avoid explosive missteps, we rely on data and statistical reasoning to inform our thinking. However, coaxing meaning from data can be a perilous proposition. We'll consider statistical fiascos from presidential polls to airline crashes. We'll find that the average income of a school's graduates may not tell us how much the typical graduate earns. And we'll learn that flunking important medical tests may not be as fatal as it first appears. Carefully quantifying an issue allows us to put a number to our uncertainty and bet well in the gamble of life.

We face the random, uncertain, and unknowable every day of our lives. The methods of mathematical thought developed in these chapters enable us to more accurately parse our complex world and to chart a better course toward a better future. The surprising and entertaining vistas on the chancy side of life present us with a template for clear thinking. By learning to cull essential ingredients from among the many distractions, we can stare into the at-first-frightening blank face of the unknown and make it blink.

CHAPTER 1

UNBRIDLED COINCIDENCES

Likelihood, Lady Luck, and Lady Love

*Chance, too, which seems to rush along with
slack reins, is bridled and governed by law.*—Boethius

Obviously . . . Colored lights dance from spinning disco balls while
sequined servers jiggle through the crowds plying the players with
cash-loosening cocktails. All this glitter sets the tone at the Big
Wheel Casino in Las Vegas. Mounted on center stage, the giant
wheel of fortune clicks in its characteristic rhythm and then slows to
land in one of the 360 numbered slots—one for every degree of a
circle. You place your bet, then 45 guests take one spin each in turn.
If two spins coincidentally land in exactly the same slot, the casino
wins. If not, you win. Sounds like good odds—360 slots, only 45
chances to make a match. You bet the farm.

Surprise . . . You lose your shirt. In fact, the incredible coincidence of
a match will occur more than 94% of the time. Amazing coincidences happen surprisingly often.

ASK NOT IF COINCIDENCES WILL HAPPEN TO YOU . . .

It's just too eerie to be true, and yet . . .

Abraham Lincoln was elected to Congress in 1846.
John F. Kennedy was elected to Congress in 1946.

Abraham Lincoln was elected president in 1860.
John F. Kennedy was elected president in 1960.

Lincoln's secretary was named Kennedy.
Kennedy's secretary was named Lincoln.

Andrew Johnson, who succeeded Lincoln, was born in 1808.
Lyndon Johnson, who succeeded Kennedy, was born in 1908.

John Wilkes Booth, who assassinated Lincoln, was born in 1839.
Lee Harvey Oswald, who assassinated Kennedy, was born in 1939.

A week before Lincoln was shot, he was in Monroe, Maryland.
A week before Kennedy was shot, he was in . . . well, you get
the idea.

Unbelievable! What are the chances? What's going on? Is there
a cosmic conspiracy? This can't just be dumb luck—or can it?

In fact, coincidences *do* happen, and when they do, we take note.
Any particular coincidence is indeed an extremely rare event; however, as we'll soon see, what is rarer yet is for us to experience *no*
coincidences at all. Basically, the moral is to expect the unexpected.

DEALING WITH COINCIDENCES

Here's something you really must try. You can do it alone, but it's
more fun with two people. Take two decks of 52 playing cards, and
give one deck to your friend. Shuffle each deck as many times as you
wish. Then, at the same time, you both turn over the top cards from

the decks and place those two cards face up on a table. Then repeat, simultaneously turning over the top cards of the decks and placing them on the piles of face-up cards, until you've gone through the decks. Will there be a match? That is, will there be a moment when you and your friend turn over the exact same card at the exact same time—for example, you each flip over the ace of spades or you each flip over the jack of diamonds? Suit yourself and try reshuffling and going through the decks several times. Surprisingly, it turns out that we should experience at least one "amazing" match approximately two out of every three times we go through the decks.

Coincidences surprise us because our intuition about the likelihood of an event is often wildly inaccurate. We will not delve here into the mathematical details behind the surprisingly high chance for a match in the card experiment. Simply put, the underlying principle is that if we have many opportunities to witness some rare event, then it is extremely likely that eventually we'll see it. In the case of the cards, each individual flip is unlikely to yield a match, but with 52 opportunities, chances are pretty good that some match will occur. Knowing the odds makes the matches less amazing but may make our expectations about life more accurate.

PRESIDENTIAL PARALLELISMS

Let's take a closer look at the amazing Lincoln-Kennedy parallels. Should we truly be amazed? Whether this uncanny string of parallels is amazing or not amazing is really the question. Certainly the parallels are striking curiosities, but what we would like to know is whether we should expect such coincidences by random chance or if the existence of such parallels is an eerie, supernatural message from the great beyond.

To put this situation in perspective, let's note first that assassinations of charismatic presidents tend to attract some attention. Literally hundreds of thousands, perhaps millions, of facts (and myths) have been amassed (and made up) about Lincoln and Kennedy— their lives, presidencies, and assassinations. Lincoln and Kennedy are not average citizens. The pile of minutiae through which to forage for possible coincidences is truly immense. Just think about the seemingly endless collection of life data that we can consider. How many people are associated with Lincoln and Kennedy? How many dates are associated with their lives and with the people connected to their lives? How likely is it that there would be *no* coincidences of dates and names among this blizzard of possibilities? The likelihood that there would be no coincidences is essentially zero.

Every event in the life of a president involves a date. If both Lincoln and Kennedy experienced the same life event, then we have a pair of dates to consider. Given that their lives were approximately one hundred years apart, among their thousands of life events and date pairs, it is essentially certain that some of the dates would differ by exactly one hundred years. Any particular coincidence, such as the dates of the two future presidents' election to Congress, is unexpected, but among thousands of possibilities, we would certainly expect some dates to correspond. So parallels of some dates are *expected* coincidences. If *huge* numbers of life dates for Lincoln and Kennedy were exactly one hundred years apart, then we would have to think carefully about the possibility that famous people are reincarnated every hundred years like clockwork. (Maybe we could consult Shirley MacLaine.)

But the concept of centennial recycling doesn't get much support when we take a hard look. (Sorry, Shirley.) Most of the Lincoln-Kennedy dates we could consider do not coincide—for example, year of birth; year of graduation; year of marriage; year of death; year of birth of mother, father, brothers, sisters, children, cousins, grandchildren; and so on. Every family event, every professional event, every national event, every life milestone offers a potential coincidence in year. How many coincidences should we expect? That depends on how many events we consider.

One of the keys to putting coincidences in perspective is to realize that we usually haven't decided what type of coincidence we are seeking before we happen to witness it. In this case, no one dictated that the date parallels had to be pairs of *years*. We could, alternatively, look at days of the month instead of years. If there were coincidences between days of the month—let's say March 23—for some of these events, rather than between years, then the events taking place on March 23 would be the ones that would have been listed as remarkable coincidences. Given that there are only 366 possibilities for calendar dates and over four score and seven thousand potential commonalities, we know that those coincidences are out there still to be tapped by psychics, presidential biographers, and the *National Enquirer*.

We see a similar scenario regarding individuals' names. How many people are associated with any person, and how many thousands of people are associated with prominent presidents? Answer: Many. With thousands of people to choose from, some coincidences are bound to occur. Finding coincidences among millions of possibilities is an entirely different proposition from looking at just one question. If Lincoln and Kennedy were assassinated and we said, "I don't want to know anything except the names of their secretaries," then we could be impressed by the coincidence of names. But if their secretaries had been named Smith and Woods, the observation about their names would not have made our "amazing" list. On the other hand, if the secretaries' names had been Smith and Wesson, they would have been on the list as coincidental gun names—again, certainly a long *shot*, but it could have happened.

The coincidences about Lincoln and Kennedy are notable and known because Lincoln and Kennedy are famous. However, if we took any two ordinary Joes or Janes and delved as deeply into their lives as historians and journalists have delved into the lives of Lincoln and Kennedy, we would find amazing coincidences there too. Coincidences do not arise because the people are prominent. They arise when we ask so many questions that the vast numbers of opportunities make the chance for coincidences overwhelming. The Lincoln-Kennedy similarities don't come from the grassy knolls of covert cosmic conspiracies, but rather from the mathematical certainty of coincidences.

MAKING WAVES IN THE GENE POOL OF TWINS

What fascinates us about identical twins is that they are so much alike. For one thing, they look alike. Their genes are identical. To what extent, however, do they share personality traits? This question intrigues psychologists who are trying to tease apart the contributions to personality made by genetics from those caused by upbringing—nature versus nurture. The best method for studying such influences would be to take the next one thousand pairs of identical twins born in the United States, separate each pair at birth, and have each child brought up by a different family; later we would measure the similarities and differences among the pairs. But in our selfish society it's not easy to locate many parents of newborn twins who are willing to give up their children for the twenty-year experiment. (We do note, however, that it's less difficult to find donors if the twins are in the terrible twos.) So psychologists find themselves on the horns of a dilemma—namely, how to get the data without resorting to the ethically questionable experiment of kidnapping thousands of infants.

This dilemma forces psychologists to rely on the method of choice in much of life—dumb luck. Once in a great while, identical twins are actually separated at birth and brought up in different households where there is no interaction between the twins. These pairs are as rare as steak tartare. So when they are found, psychologists squeal in delight and rush in with an orgy of questionnaires.

A famous example of identical twins separated at birth resulted in incredible findings. These two siblings, who discovered each other's existence at the age of 39, turned out to have uncanny commonalities:

- Both were named James.

- Both had a white metal bench around a tree in their yard.

- Both had a first wife named Linda and a second wife named Betty.

- Both had a dog named Toy.

- Both drove a Chevrolet.

- Both chain-smoked Salems and drank Miller Lite beer.

- Both had a son—one named James Alan, the other James Allen.

- Both appeared on *The Tonight Show Starring Johnny Carson* on the very same night!

Proof of the influence of genetics on personality? Perhaps. But to interpret these results as supporting the idea of genetic influence on personality, we need more information, just as we did in the case of the Lincoln-Kennedy coincidences. How many questions were asked? Take two random people off the street of about the same age who were brought up in circumstances similar to those of the twins. Ask them thousands of questions about their lives, their interests, their brand preferences, their families, and their innermost desires. If we ask these two random people thousands of questions, how many coincidences would we see? About the same as occurred with the twins? If so, then the genetic explanation may not be the right one. On the other hand, most people with more than one child are quite certain that genetics play a large role in personality.

How can we interpret the identical-twins data persuasively? Coincidences in names of spouses and sons are probably poor examples of genetic influence. Somehow it strains credibility to think that our desire to take a bride with the common name of Betty comes from inside our genes. And the coincidence of the twins' own names being the same is clearly an example of random luck, since they didn't name themselves. Beer-brand preferences may or may not be meaningful examples. Taste may well be partly determined by the structure of taste-bud cells, but on the other hand there are only a limited number of brands of beer, so coincidences in beer preference are not that unlikely with two random people. (This taste bud is for you . . . but for others as well.) We would have to compare the coincidences at hand with the coincidences we would expect to see generated by randomness alone.

Coincidences in twin studies may be significant or not. But what is clear is that interpreting such studies requires far more information than just the list of coincidences.

MONKEY BUSINESS—HOW TO WRITE A PULITZER PRIZE-WINNING PLAY

If some process is generating things truly randomly, then that procedure is unable to distinguish between surprising outcomes and commonplace outcomes—thus with the genuinely random we should experience both. Every particular poker hand we are dealt is exactly as unlikely as any other hand. So logically, our friends should be just as amazed to hear about the day we were dealt a 3♣, 5♦, 8♣, 9♠, and Q♠ as they are to hear about the day we unfolded a royal flush in spades. Since every possible hand is as likely as any other, if we are dealt sufficiently many poker hands, we should expect eventually—though not necessarily in our lifetime—to be dealt a royal flush. Monkeys at typewriters make this point more dramatically.

Suppose a roomful of monkeys were placed in a room full of word processors, and the monkeys started randomly pecking letters on the keyboards and continued *forever*. Eventually one would type *Hamlet* in its entirety, without a single error. Why? Because there is some chance, although fantastically tiny, that some monkey would randomly strike the exact sequence of keystrokes to produce *Hamlet* purely by accident. Amid all the crap that the monkeys would produce, we would occasionally find a literary pearl. *Hamlet* happens.

In fact, those monkeys would also type out some of Shakespeare's pre-edited versions of *Hamlet*. For example, some versions would render Hamlet's "To be or not to be? That is the question" soliloquy in the words of Shakespeare's original draft, "Two bee or not two bee? That is the query that buzzeth."

After all, forever is a long, long time. So if those monkeys kept pecking away at random, the unlikely striking of the correctly ordered 31,281 words that make up *Hamlet* would happen by and by. Wouldn't it be exasperating to get all the way to the end and then see "The Enb" and know that that poor monkey has to start all over? But with forever comes unending patience—and plenty of drafts, including the one with the typo in the very last word. In fact, those monkeys would also type out every other book ever written, every book to be written, and every variation on them all. Just two monkeys (named Ed and Mike), for example, were enough to write this book.

The random typing of *Hamlet* is such an unlikely event that in reality it will not happen in our lifetime or even during the entire history of the universe. So the hypothetical random act of typing *Hamlet* is an abstract idea. In fact, though, incredibly unlikely occurrences do happen all the time. The exact timing necessary for each of us to come into existence and the exact arrangement of our brain cells and our life stories all involve random events that are utterly unlikely and cannot be duplicated. Viewed in this manner, each of our lives resembles a monkey's typed version of *Hamlet*.

GOOD KARMA, BAD THINKING

Some people believe they enjoy incredible "parking karma": they can drive up to a theater just minutes before showtime and always find a parking spot right by the entrance. Others feel that they suffer from a malady known as bad-checkout-line karma: they always pick the slowest line at the grocery store or the slowest line at the highway toll plaza. A variation can be found at stoplights. Why do some of us hit all the red lights when we're in a rush? Are there impish auto angels straddling our hood ornaments who clear parking places as needed for deserving drivers and turn the lights red for the rest of us?

As much as we might want to believe in that fanciful image, the actual explanation is much more mundane and karma-free. In reality, we all choose the slow lines about as often as anyone else, and we are all lucky sometimes. What differs is which kind of event we remember.

Suppose, for example, we are sitting in a long line watching all the other lines flow through the tollbooths like rivers while driver after driver in front of us parks in the lane and feels around under the seats in search of that needed nickel. During these long delays we have plenty of time to sit, think, and, in the case of many of us, mutter inappropriate expletives. That leisure gives our brains time to burn that unpleasant experience into our memory. That's where the bad karma comes in. But when we zoom through the line without any hesitation at all, it's so fast that most of us don't even notice and thus miss the opportunity to celebrate the good luck we've just enjoyed. Those of us who tend toward pessimism or self-pity remember the wait times and feel caught in the grip of bad-checkout-line karma; the lucky few who are optimistic remember the triumphant moments and bask in the illusion of good karma. The bottom line is that there *is* an amazing parking spot with your name on it, right by the door . . . every once in a while.

Bad-elevator karma, however, is a different story. Suppose we work on the third floor of a twenty-story building and regularly take the elevator to the seventeenth floor to flirt with an attractive executive. It's annoying that the first elevator that arrives at our floor is almost always going down. Bad karma? Not really; it's simply that there are more floors above us than below, so that's where the elevators are more likely to be. When an elevator gets to the third floor, it typically will be coming from one of the many higher floors, heading down to drop off its passengers in the lobby. So this bad-elevator karma is neither an illusion nor karma.

A little solid reasoning is usually more reliable than riding the ups and downs of mystical influences. It's remarkable how many seemingly inexplicable phenomena can be moved into the "Aha! I see it now!" category when you do the math. Some people may see the trade-off as undesirable: they feel they're giving up something mysterious, inexplicable, and wonderful in exchange for simple, boring, everyday reasoning. Not to worry—there's plenty of real mystery left in the world.

PICKING A WINNER—STOCK EXPERTS, ASTROLOGERS, MONKEYS, AND DARTS

As long as we're considering karma and the literary abilities of monkeys, it seems natural to discuss the wisdom of stock market gurus. Many predict the future of the stock market—Ph.D.s in finance, business tycoons, astrologers, monkeys, and darts. So how do the astrologers and monkeys fare? We'll start with a scenario of a financial advisor who really gets it right.

Monday morning when we turn on our computer, we are deluged with many unsolicited e-mails about products that make us blush. One unsolicited message, however, is from the investment advisors at E. F. Nuttin'. We open this message and delete the others. The message is short: "At the end of this week, Dell will close up." We, of course, dismiss and delete this spam.

The following Monday morning, we repeat the ritual of deleting those e-mail opportunities to improve our private parts. Once again we find an E. F. Nuttin' message: "Last week we correctly predicted that Dell would close up. By the end of this week, Dell will close down." We again hit the delete key, but then we find ourselves looking up the stock prices to see if last week's prediction was indeed correct. It was.

The following Monday morning, we are no longer surprised to see an unsolicited message from E. F. Nuttin'. The message is again short: "We correctly predicted the last two weeks' performance of Dell. At the end of this week, Dell will close up." We verify their claim of predictive accuracy and begin to look forward to next week's advice.

This ritual is repeated for nine weeks in a row, with E. F. Nuttin' compiling a perfect predictive record. If we had taken their advice from the beginning, our fortunes would have soared. On the tenth Monday, we are surprised to discover that the ritual has changed. The E. F. Nuttin' message this time reads: "We have correctly predicted the future performance of Dell for nine consecutive weeks. Please remit by return e-mail the sum of $1,000 to receive next week's advice. As an added incentive, we offer a money-back guarantee: we will provide a full refund of the $1,000 fee if our prediction is not correct."

Obviously. We excitedly wire the rather modest $1,000 fee, which we feel is the best investment we have made in our young lives. We eagerly await the prediction, which dutifully arrives by return e-mail. Our investment strategy for the next week is clear—buy or sell short as recommended.

Surprise. E. F. Nuttin' has perpetrated a systematic scam, and their advice is worthless. Let's see how they produced such impressive accuracy.

On week one, E. F. Nuttin' sent out 1,024 e-mails to 1,024 people. To 512 people they wrote that Dell would go up in the coming week, and to 512 they wrote that Dell would go down. At week two, they sent out 512 e-mails—to those for whom they were right the first week. To 256 people they predicted that Dell would go up in the week to come, and to 256 they predicted that Dell would go down. At this stage, 256 people had received correct predictions of the future for two consecutive weeks. At week three, they sent 256 e-mails—128 said up, 128 said down. At week four, they sent 128 e-mails—64 ups and 64 downs. Week five went to 64 people; week six to 32; week seven to 16; week eight to 8; week nine to 4. (*Figure 1.1.*)

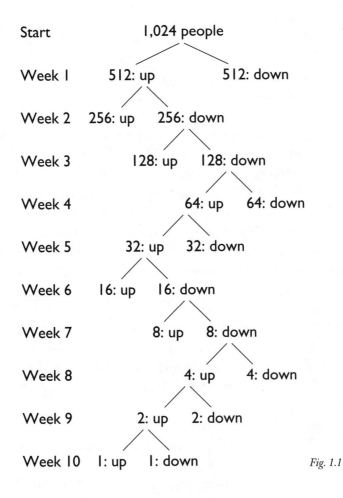

Fig. 1.1

By week ten, two people have received correct predictions about the future for nine weeks in a row without error! Now E. F. Nuttin' asks them each for $1,000 for next week's pick—with a money-back guarantee. From the perspective of the two receivers of consistently correct predictions, the advisors at E. F. Nuttin' clearly know what they're doing. Each of the two investors remits $1,000. E. F. Nuttin' then sends one person an "up" e-mail and the other a "down" e-mail as the next week's prediction. The following week E. F. Nuttin' refunds $1,000 to the unlucky investor for whom the prediction was wrong and pockets the other $1,000—"earned" through their amazing stock-prediction skills.

This ploy is simple, effective, and, no doubt, illegal. However, the exact same principle is at work inadvertently all the time. Consider the thousands of people out there who express their opinions about the future of the stock market. Of course, they all think they are making their predictions based on sound reasoning. Some look at cost-to-earnings ratios (whatever that means), some look at patterns on graphs, some study the companies and know their specific market possibilities, and some use the alignment of the planets to predict the financial future.

Of course, most people would scoff, "The stock astrologer is a quack." But the fact is that a certain number of astrologers, clairvoyants, and even stockbrokers will be right in their predictions. Why? Because there are so many of them making random guesses that some of them, purely by accident, will be uncannily correct—real life will accidentally coincide with a prediction. This is exactly the same situation as the e-mail scam, except instead of many e-mails that offer different predictions, there are many predictors, each making one prediction. Of course, some predictors may have legitimate reasons for their predictions. The challenge for investors is to tell the difference between those few and the many who would do just as well by throwing darts. We recommend darts and monkeys—they are sometimes right and have the advantage that they don't charge a commission. In fact, you can hire a monkey for peanuts.

THE LOTTERY

Losing is always annoying; hence, lotteries are annoying. In a typical state lottery, six numbers between 1 and 50 are chosen. If we are foolish enough to invest in a ticket, then after the drawing we kick ourselves for our number choices. We see the winning numbers and say, "Oh, gee, I know those numbers! One of them is in the middle of my cell phone number, another is the birthdate of my aunt Edna. Why didn't I choose those six instead of the loser numbers I selected?" It turns out that there are about 16 million possibilities for choosing six numbers, so perhaps we shouldn't be too hard on ourselves when we lose. In fact, with 16 million choices, the person who does win is amazingly lucky. That person made a 1-in-16-million choice. Amazing.

So why aren't we astounded every time there is a winner? That winner is certainly astounded with his heart racing (as he runs from the IRS). We are not astounded because we know that when the jackpot gets above the paltry few million dollars where it starts and moves into the 50-million-dollar range, which we can regard as real money, then tens of millions of tickets are sold. So if tens of millions of people try something that has a 1-in-16-million chance, our intuition correctly tells us that someone is likely to hit the jackpot. While some individual winners enshrine their lucky socks and wear their lucky underwear forever (even though they can now afford as much clean underwear as they desire), we know that the amazing coincidence of having *some* winner is not amazing at all.

In any case, the point is that if enough attempts are made, even very unlikely events will happen. If you blindfold yourself and repeatedly throw darts in the general direction of a dartboard, you are extremely likely to hit the bull's eye eventually. The moral in the case of the lottery is that if you want to improve your chance of winning, you should buy a few million tickets rather than just one. If you want to be certain to win and have 16 million dollars to invest, just buy every possible set of numbers.

SURPRISE BIRTHDAY PARTIES

While people are different in many ways, we all started off the same—we were born. Every year has 365 days, or 366 if it leaps. Each person was born on one of those 366 days. We all celebrate our birthday (or at least until we get old enough to know better). And when our special day arrives, we want to be pampered like a king or queen—so it is slightly annoying when we have to share our birthday spotlight with some usurper. Unfortunately and somewhat surprisingly, chances are that even in rather small groups there will be people who do not have their special day to themselves. Coincidences of birthdays are far more common than many people imagine. On the bright side, an accurate understanding on your part of the surprisingly high chances for birthday coincidences can lead to profitable gaming opportunities with those whose intuition of coincidences is . . . well, more youthful and naïve than yours.

The next time you find yourself in a room with 45 people, brag about your power to predict amazing coincidences and announce that you sense in this small group at least two people with the same birthday. Bet money with skeptics or with psychics who read otherwise in their tea leaves; have everyone declare his or her birthday; find an incredible match; and collect your winnings. The skeptics will then want your stock picks for the next year, and the psychics will want your insights on which Hollywood marriages will soon hit the skids.

Why is it that you can be so confident of finding this "amazing" birthday coincidence with so few people? At first glance, it seems quite unlikely that two people out of 45 would have the same birthday—after all, there are 366 days to choose from. It turns out, however, that in a random group of this size there is about a 95% chance of a birthday overlap.

To demystify this reality, let's begin by gaining some experience. The best way to gain experience would be to take forty-five 366-sided dice (one side for each day of the year) and roll the dice and see how often we get at least one pair. About 19 out of 20 times, in fact, we would see at least one match! Of course, since very few people carry around forty-five 366-sided dice so as to be ready for this ultimate crapshoot, we will have to resort to our own reasoning.

Our goal is to answer the question "What is the chance that among 45 random people two or more of them have the same birthday?" Looking at the question the other way around, we could ask, "What is the chance that among 45 people *none* of them has the same birthday as anyone else?" If we can answer either question, then we will immediately be able to answer the other. And it turns out that it is easier to find the chance that no one shares a birthday—that is, the chance that all these people have different birthdays.

Let's just imagine a lineup of 45 people. We start with the first person and ask, "What is your birthday?" We record the answer and move on to the second person. To *avoid* a match, that second person must have a birthday different from the first person's. The chance of having a *different* birthday is very good—365/366 or 99.7%. To avoid a match, the third person must miss both of the first *two* birthdays. The chance of that is 364/366 or 99.4%, since there are 364 unused birthdays at that person's disposal. Each subsequent person must miss increasingly many birthdays. As we approach the end of the line, the 44th person must have a birthday that misses 43 previous birthdays to avoid a match. The chance of that is 323/366 or 88.3%—still a good chance. The last person has a chance of 322/366 or 87.9% to miss the previous 44 birthdays. Each individual's chance of missing all of the others is therefore pretty good—but that's not what we want to know. What is the chance that *all* of these people miss the birthdays of all of the others? That chance turns out to be rather remote.

To compute the chance that all of those people have different birthdays, we multiply together the *individual* chances that we came up with above: 365/366 × 364/366 × . . . × 323/366 × 322/366. Notice that each of those numbers is a fraction less than 1, and if we multiply a list of fractions less than 1, the product is extremely small. (If it's not obvious why the product is so small, think about cutting a pie. If you cut half a pie in half again and then cut one of *those* pieces in half, you've got an eighth of a pie: $\frac{1}{2} \times \frac{1}{2} \times \frac{1}{2} = \frac{1}{8}$.) Roughly speaking, all those fractions average to around .935, so an estimate of that product of the 44 fractions is about $.935^{44} \approx 0.05$ or 5%. That is, there's about a 5% chance that *all* the birthdays are different. So the alternative, that there is at least one match, is about 95%. This little bit of simple arithmetic has revealed that even in relatively small groups, some people are likely to have to share the spotlight on their birthday. This coincidence is nearly certain!

MAGIC

The fact that coincidences frequently occur confounds our intuition, because each individual coincidence is truly a rare event. The striking and unexpected experience of a rare coincidence sometimes tempts us to give magical significance to random happenings. Attaching meaning to random occurrences is a great source of all kinds of mystical foolishness. In reality, each individual's life is actually a long sequence of incredibly unlikely events all strung together—our lives' paths are filled with magic. Consider for example, the incredible chain of events that brought you to this book and to this page. Think about all those unlikely circumstances that led you to consider the subject of coincidences as you read this very sentence—where we're talking about coincidences! What an amazing coincidence!

CHAPTER 2

CHAOS REIGNS

Why We Can't Predict the Future

God has put a secret art into the forces of Nature so as to enable it to fashion itself out of chaos into a perfect world system.—Immanuel Kant

Obviously . . . All sound in the computer lab came to a crashing halt. We could almost hear the musical theme of *High Noon* playing in our heads as all eyes turned to the two skinny programmers standing on opposite sides of the lab. They shot angry looks at each other from behind their thick black-framed glasses. Each programmer claimed a different numerical answer to the same calculation, and each was ready to duel to the digital death. When given the signal, they drew their calculators from their pocket-protector holsters, entered the same decimal number (0.37), and punched the same sequence of three keys—the key for squaring (x^2), the minus sign ($-$), then the number 2 (2). Of course, their calculators displayed the same result. Then they punched the same three keys again. They kept punching those three keys, thirty times in succession. Both calculators were working perfectly. Obviously, these two calculators, these symbols of stability and certitude, would produce the same answer when all thirty rounds were done.

Surprise ... The two calculators displayed wildly different outputs— and they were *both* wrong. All was not well at the O-K Computer Lab. Chaos reigns.

We saw in the last chapter that amazing coincidences *do* happen by random chance alone. Coincidences are unexpected similarities—that is, unexpected convergences. In this chapter we explore the opposite phenomenon, unexpected differences—in other words, unexpected divergences. Boisterous butterflies, specious spreadsheets, crazy calculators, peculiar pendulums, and bouncing balls will all baffle our intuition with counterintuitive chaos. Minute, insignificant changes in our world today become magnified over time and dramatically alter our future. We'll begin where many discussions of chaos do, with the tragic tale of a butterfly from Brazil whose careless flapping has caused more tornadoes in Kansas than all the future sequels to *The Wizard of Oz* combined.

THE INFAMOUS BUTTERFLY FROM BRAZIL

The parable begins in a Brazilian rainforest moist with dew. The main character of our tale is a beautiful, delicate butterfly that lazily but gracefully beats its fragile wings. The air around those wings stirs ever so slightly, but that slight air current slightly deflects a slightly larger current of air. That air mass, in turn, influences another volume of air that is larger still.

Over time, that slight wing movement ripples outward step by step to influence larger and larger air masses moving with ever-increasing power. Soon our butterfly's gentle stretch is causing thunderheads to form, and several severe storms are swirling over the globe. These storms move enormous volumes of air until we see the ominous signs of tornadoes touching down in Kansas. The sky darkens, the winds race, and it begins raining cats and—well, let's just say that it's not a great day for poor Toto.

If only that thoughtless butterfly in Brazil had been more sensitive and curbed its fleeting desire to flutter, Kansas would have enjoyed a calm, warm, sunny day with Toto's tiny feet firmly planted on the ground. That fateful flap changed the course of history, and the catastrophe-creating butterfly inspired the term *butterfly effect*, which refers to the fact that slight changes today cause enormous changes in the future.

Looked at from another point of view, the existence of the butterfly effect raises the intriguing question of whether we could become the butterfly. That is, could we learn how to *intentionally* make the subtle changes that would affect future weather in dramatic but controllable ways? Perhaps we could seed clouds judiciously to create crop-saving rain in times of drought, for example, or to make hurricanes that would attack enemies in times of war.

Whether or not control of some aspects of the weather is in our future, the butterfly tale suggests that we will never know how to predict the weather even a month or two into the future. Countless tiny influences that we could never hope to measure are at work, altering the present in countless tiny ways that will be magnified to lead to enormous changes in a month. Computers may become thousands of times more powerful, cellular phones may be implanted in our ears, and world peace may reign, but we'll never know in April whether our outdoor wedding will be rained out next June 16. There will always be some insensitive butterfly carelessly flapping and unintentionally wreaking havoc on our nuptials.

The fact that we cannot predict the future is not breaking news, of course—just ask any stockbroker who was pushing Enron (if you can find one not behind bars). We all know that in real life tiny differences in time and place lead to dramatic differences in our lives. It is a platitude that our choices at the many forks in the road of life change our destinies forever. The big news comes when we develop a mathematical reflection of this simple principle. The consequences of this mathematical development cause us to rethink the stability of everything from pendulums to populations.

MATH FROM LIFE

Mathematics is born by abstracting concepts from life experiences or observations. The butterfly effect, for example, gave birth to mathematical chaos, which refers to simple mathematical processes in which minute changes early on lead to dramatic differences later. The word *chaos* can become a bit, well, chaotic, so we need to be clear about its various meanings:

1. *Chaos* in everyday English: The dictionary defines the word as "a state of utter confusion or disorder; a total lack of organization or order."
2. *Chaos* in the actual world: The second meaning is illustrated by the butterfly and the weather. *Chaos* here refers to the phenomenon in which a slight change in the situation at one moment has only a small effect at first but is then magnified with each subsequent step in the process. The eventual effect is a vast, but theoretically predictable, influence on the future. In other words, the butterfly does not cause *random* weather, it causes *different* weather. The phrase associated with this idea is "sensitivity to initial conditions."
3. Mathematical *chaos*: This sense of the word has to do with repeated mathematical processes that have a sensitivity-to-initial-conditions property. That is, mathematical chaos is an extrapolation of the real-world butterfly effect. It is not the same as chaos in ordinary English, because it possesses neither randomness nor uncertainty. Instead, a mathematical process that exhibits chaos, although completely accurate and deterministic, diverges very quickly from the results obtained with even slightly different initial starting points.

It is this third sense of the word, the mathematical side of chaos, that we want to explore in the remainder of this chapter.

CHAOTIC SQUARING

Repeating a simple procedure leads to complexity. To illustrate this surprising fact, we will show what happens when we repeatedly carry out one of the simplest mathematical processes imaginable: taking a number, multiplying it by itself (that is, squaring it), and then subtracting 2. In order to build insight into this seemingly predictable process, we begin with some simple numerical experiments.

If we start with 0, then 0 squared is 0, minus 2 equals −2. Next, −2 squared is 4, minus 2 is 2. Now, 2 squared is 4, minus 2 is 2. At this point the process stabilizes: No matter how many times we repeat the procedure of squaring and subtracting 2, we'll always come out with 2. If we now start with 1, then 1 squared minus 2 is −1; next, −1 squared is 1, minus 2 is −1. Again the process stabilizes; in this case the output is −1 forever. Finally, let's start the process with 3. First we note that 3 squared is 9, minus 2 yields 7. Next, 7 squared is 49, minus 2 is 47. Next, 47 squared is 2,209, minus 2 equals 2,207. We see that if we start with 3, the generated values get larger and larger. In each of the three examples, one point is clear: It does not matter who is doing the squaring and subtracting—any two people who can correctly multiply and subtract will generate the exact same list of values. Not much chaos here. But let's not give up quite yet.

Let's see what happens if we start the process with a decimal number such as 0.5. Using a spreadsheet computer program, such as Excel, provides us with a simple method for repeatedly applying a mathematical process, and it allows us to easily perform experiments on our own. So, using Excel, place a number—in this case, 0.5—in the top of a column, say in the A1 cell. In the next cell down—that is, cell A2—enter the formula "=A1^2 − 2." Highlight the A2 cell and the column below it; do not include the A1 cell in the highlighting. When you select the "Fill down" command, values obtained from the iterative process—that is, a process involving repetition—will fill the column's cells. The moment you enter a different value in the top cell of the column, Excel changes all the lower cells automatically.

Starting with 0.5, we simply put that baby in the top of the column and let Excel work its digital magic. We can extend the column to reveal thousands of future iterations. The first column of Figure 2.1 displays the first few dozen entries. What do we see in that ocean of numbers?

COLUMN A	COLUMN B
0.5	0.50001
−1.75	−1.74999
1.0625	1.062465
−0.87109375	−0.871168124
−1.241195679	−1.241066099
−0.459433287	−0.459754937
−1.788921055	−1.788625398
1.20023854	1.199180813
−0.559427448	−0.561965379
−1.687040931	−1.684194913
0.846107103	0.836512505
−1.284102771	−1.300246828
−0.351080073	−0.309358186
−1.876742782	−1.904297513
1.52216347	1.626349018
0.316981629	0.645011127
−1.899522647	−1.583960646
1.608186286	0.508931328
0.58626313	−1.740988903
−1.656295543	1.031042361
0.743314925	−0.936951651
−1.447482922	−1.122121604
0.09520681	−0.740843105
−1.990935663	−1.451151493
1.963824815	0.105840656
1.856607905	−1.988797755
1.446992912	1.955316512
0.093788487	1.823262662
−1.99120372	1.324286736
1.964892253	−0.24626464
1.860801568	−1.939353727
1.462582474	1.761092878
0.139147493	1.101448125
−1.980637975	−0.786812028
1.922926789	−1.380926832
1.697647434	−0.093041085
0.882006811	−1.991343357
−1.222063986	1.965448364
−0.506559614	1.86298727
−1.743397358	1.470721568
1.039434347	0.16302193
−0.919576239	−1.97342385
−1.154379541	1.894401693
−0.667407875	1.588757773
−1.554566728	0.524151261
0.416677713	−1.725265455
−1.826379683	0.976540891
1.335662748	−1.046367888
−0.216005025	−0.905114242
−1.953341829	−1.180768208
1.815544302	−0.605786439
1.296201113	−1.633022791

Fig. 2.1

What we see is chaos. In fact, the result is even more chaotic than it looks. To expose the hidden chaos, let's perform the same procedure of squaring and then subtracting 2 in the B column, but this time with a number that is nearly identical to .5—say, .50001. The results are given in the second column of Figure 2.1. We make the startling observation that after a few steps the numbers in the two columns differ dramatically. So a nearly insignificant difference in initial values very quickly yields totally different outcomes.

We all believe, properly, that electronic devices such as calculators and computers do not make arithmetical mistakes—that is, if they add two numbers, they will always return the same correct answer. Performing calculations correctly is what calculators and computers were born and bred to do. Given this understanding, let's consider an example that may prove to be somewhat disturbing. Let's return to our Excel spreadsheet, and as before, let's enter 0.5 in the A1 cell, type the formula "= A1^2 − 2" in the A2 cell, and "fill down" the values in the A column (*Figure 2.2*).

COLUMN A	COLUMN B
0.5	0.5
−1.75	−1.75
1.0625	1.0625
−0.87109375	−0.87109375
−1.24119568	−1.24119568
−0.45943329	−0.45943329
−1.78892105	−1.78892105
1.20023854	1.20023854
−0.55942745	−0.55942745
−1.68704093	−1.68704093
0.8461071	0.8461071
−1.28410277	−1.28410277
−0.35108007	−0.35108007
−1.87674278	−1.87674278
1.52216347	1.52216347
0.31698163	0.31698163
−1.89952265	−1.89952265
1.60818629	1.60818629
0.58626313	0.58626313
−1.65629554	−1.65629554
0.74331492	0.74331492
−1.44748292	−1.44748292
0.09520681	0.09520681
−1.99093566	−1.99093566
1.96382482	1.96382482
1.8566079	1.8566079
1.44699291	1.44699291
0.09378849	0.09378849
−1.99120372	−1.99120372

1.96489225	1.96489225
1.86080157	1.86080157
1.46258247	1.46258247
0.13914749	0.13914749
−1.98063798	−1.98063798
1.92292679	1.92292679
1.69764743	1.69764743
0.88200681	0.88200681
−1.22206399	−1.22206399
−0.50655961	−0.50655961
−1.74339736	−1.74339736
1.03943435	1.03943435
−0.91957624	−0.91957624
−1.15437954	−1.15437954
−0.66740787	−0.66740787
−1.55456673	−1.55456673
0.41667771	0.41667771
−1.82637968	−1.82637968
1.33566275	1.33566275
−0.21600502	−0.21600502
−1.95334183	−1.95334183
1.8155443	1.8155443
1.29620111	1.29620111

Fig. 2.2

Now let's repeat the exact same procedure in the B column. We mean literally the exact same thing—we start with 0.5 in the B1 cell, enter the formula "$= B1^2 − 2$" in the B2 cell, and "fill down" the column. No surprise: Excel generates the same answer in each B cell as it did in the corresponding A cell.

Notice again that the two columns are identical. Both are repeating the same process of squaring and subtracting 2; both start with 0.5 as the initial value. We now do something a bit bizarre. We select a row—let's say the twelfth one—and notice that the entry in A12 is equal to that of B12, as are they all. Let's now simply *retype* all the digits of the decimal number that appears in B12—that is, we type in the B12 cell the exact number that we saw in B12 to begin with and that we still see in cell A12. This activity appears utterly pointless, since it seems as though we're not doing anything. Still, let's do it.

We copy every single digit, and Excel automatically completes the rest of the list in the B column based on this value in the twelfth cell. Since we retyped the same value, we expect that all the future values will be identical.

Surprise. Although the two columns remain almost the same at first, they soon become wildly different (*Figure 2.3*).

COLUMN A	COLUMN B
0.5	0.5
−1.75	−1.75
1.0625	1.0625
−0.87109375	−0.87109375
−1.241195679	−1.241195679
−0.459433287	−0.459433287
−1.788921055	−1.788921055
1.20023854	1.20023854
−0.559427448	−0.559427448
−1.687040931	−1.687040931
0.846107103	0.846107103
−1.284102771	−1.284102771
−0.351080073	−0.351080074
−1.876742782	−1.876742782
1.52216347	1.52216347
0.316981629	0.316981629
−1.899522647	−1.899522647
1.608186286	1.608186287
0.58626313	0.586263134
−1.656295543	−1.656295538
0.743314925	0.743314909
−1.447482922	−1.447482946
0.09520681	0.095206878
−1.990935663	−1.99093565
1.963824815	1.963824764
1.856607905	1.856607704
1.446992912	1.446992167
0.093788487	0.09378633
−1.99120372	−1.991204124
1.964892253	1.964893865
1.860801568	1.860807899
1.462582474	1.462606037
0.139147493	0.139216419
−1.980637975	−1.980618789
1.922926789	1.922850786
1.697647434	1.697355144
0.882006811	0.881014484
−1.222063986	−1.223813479
−0.506559614	−0.502280569
−1.743397358	−1.74771423
1.039434347	1.054505031
−0.919576239	−0.88801914
−1.154379541	−1.211422008
−0.667407875	−0.532456719
−1.554566728	−1.716489842
0.416677713	0.946337377
−1.826379683	−1.104445568
1.335662748	−0.780199986
−0.216005025	−1.391287981
−1.953341829	−0.064317753
1.815544302	−1.995863227

Fig. 2.3

For example, the last cell of the A column differs dramatically from the last entry in the B column. If we had room to reproduce another page or two, we'd see that the values in the A and B columns are not even remotely close to each other. We hope that you feel compelled to attempt this experiment to convince yourself that we are not fabricating this computational chaos.

What's happening here? Is our faith in computer technology misplaced? Have we found a bug in the hardware of our computer, or have those friendly folks at Microsoft let us down with their Excel software? Answer: No (and no and no).

Excel stores more digits of an answer than it displays. So although we retyped all the *displayed* digits, there exist undisplayed digits hidden from sight that make the number known by Excel in the first column slightly different from the decimal number we entered on line 12 of the second column. Of course, the numbers differ only after the tenth or so decimal digit, but our process is chaotic. So as we proceed with this chaotic procedure of repeatedly squaring and subtracting 2, we discover that quite quickly the corresponding answers become totally unrelated—all due to tiny round-off errors in digits that are so far right in the decimal number that Excel doesn't even bother to display them. In fact, the numerical news gets even worse.

BOTH COLUMNS ARE WRONG!

We cannot move beyond this surprising example of chaos without noting a very important feature about the two columns of numbers: After a few dozen steps they are both totally wrong. It's not that the first column, the column that's storing a few more digits after the decimal point, is correct while the other column is wrong. Both are utterly unrelated to the actual answer. Yes, there is an actual answer; there is no mathematical magic here. If we start with 0.5, square it, and subtract 2, there is one true answer. If that true answer is squared and we subtract 2, there is one absolutely correct answer, and so on forever. But those absolutely correct answers soon have thousands, millions, billions, and trillions of digits after the decimal point.

Just for fun, we have used a computer to write down the first eight completely correct answers to the repeated process of squaring and subtracting 2, starting with 0.5.

First iteration $= -1.75$

Second iteration $=$ 1.0625

Third iteration $= -0.87109375$

Fourth iteration $= -1.2411956787109375$

Fifth iteration $= -0.4594332871492952108383178710 9375$

Sixth iteration $= -1.78892105465919325208009812988585$
$45187613344751298427581787109375$

Seventh iteration $=$ $1.20023853980296029103616942146559$
$93282299438558451995774944565277$
$84674737691305758396790107078722$
$08740809583105146884918212890625$

Eighth iteration $= -0.55942744757165770517872120586 6417$
$74767987466759766547362611064288 4$
$33478134647384755033235341982347 0$
$73294452982645761895824039752998 1$
$91045189224960503378673326317854 9$
$84854513373905558588170169766820 5$
$77469377083473919473610713737343 7$
$58111819624900817871093 75$

Notice that the eighth answer has 256 digits past the decimal point. In order to write a number exactly, *all* of its digits must be present and accounted for. As we have seen, errors in distant digits *do* matter. But Excel has to round off numbers at some place beyond the decimal point, and those inevitable round-off errors soon propagate to make the values that Excel computes differ wildly from the correct answer. So the dozens of numbers we see in these Excel columns are completely meaningless. They are completely unrelated to the correct answers that do exist, but which we simply have no way of computing exactly. To pinpoint a number precisely, we require *all* of its digits.

Incidentally, there is a way for people who are not big Excel users to experience this numerical chaos for themselves. We simply carry out the scenario described at the beginning of this chapter, using two different brands of calculators with different decimal accuracy. If we repeat the process of squaring and subtracting 2 on both machines, we will witness chaos in action. And, as in our spreadsheet experiments, the answers on both calculators will be totally meaningless after the first few dozen repetitions.

Our Excel experiment is essentially the same one that was being performed when mathematical chaos was inadvertently discovered.

CHAOS BY ACCIDENT

Our paradigm of chaos in the world evokes images of tornadoes swirling from the butterfly effect. As fate would have it, mathematical chaos was the serendipitous discovery of a weatherman. In the 1960s, Edward N. Lorentz, a meteorologist at MIT, was producing mathematical models for weather prediction using the primitive computers of that era. He first described the weather by putting together a list of numerical data. His model used that data to generate a new list of numbers that would predict the weather for the next time increment. Then those values would be inserted back into his mathematical formulas to produce similar values for the next time increment, and so on.

On one particularly lucky day, Lorentz was running his system and generating weather forecasts when he was interrupted. He had to reenter some of the previous values and restart the process after several repetitions had already been completed. Instead of typing in all the digits of those numbers, he saved himself a bit of effort by rounding off, thinking that it couldn't possibly make any difference if he ignored the sixth or seventh digit after the decimal point. After running his model, however, he discovered that rounding off those values resulted in radically different weather predictions.

Within his mathematical weather model, Lorentz was his own butterfly from Brazil. In other words, he realized that if he plugged in values that differed only in the sixth or seventh digit after the decimal point, the mathematical model produced dramatically different values after a few repetitions. Thus he found that when performing repeated procedures, the standard practice of rounding to a certain number of significant digits results in wildly different outcomes. Lorentz realized that his system of describing weather was actually an example of a new mathematical discovery. Unintentionally, he had become the father of chaos.

PREDICTING THE FUTURE

We have been exploring *iterative systems*—processes that involve taking a value, performing some procedure to generate an answer, taking that answer as the next starting point and following the same procedure, and repeating the process again and again, generating a stream of numbers. Our first iterative system was the seemingly useless procedure of repeatedly squaring and then subtracting 2.

In reality, similar procedures model populations and even planetary locations. In each of these examples, knowing the current population or planet position allows us to apply procedures to deduce the population or planet placement at the next time step. We can repeat this process again and again. However, we now understand that after a few iterations, the results might be essentially meaningless. Thus, in our real world, iterative models of systems such as weather, population, or even planetary motion are susceptible to generating chaotic nonsense and do not produce reliable long-term predictions. The bottom line is, we now see why we'll never watch a dependable thirty-day weather forecast by our favorite wacky weatherman on the six o'clock local news.

PHYSICAL CHAOS AROUND US

Classical mechanics models the motion of moving objects. From swinging pendulums to bouncing balls to magnetic fields to fluid flows, classical physics describes precisely what will transpire. The algebraic formulas that model such physical systems are exactness personified. However, those algebraic expressions are similar to those we considered in our Excel experiments. Is it possible that the physical systems whose behavior is described by these faithful formulas actually exhibit chaotic behavior? That is, can we find physical chaos, or is this chaos business just abstract math jazz? Let's get physical.

PENDULUMS. The path of a swinging pendulum is one of the most regular patterns we know. In fact, the pendulum was for centuries the basis of clocks. If we traced the trajectory of the end of a moving pendulum, we would see a predictable path (*Figure 2.4*). Let's now consider a slightly modified object known as a *double pendulum*,

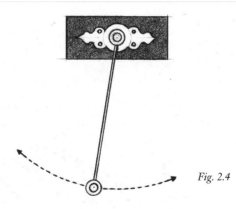

Fig. 2.4

which is simply a pendulum swinging from the end of another pendulum (*Figure 2.5*). Surely we would expect a similarly regular pattern that we could set our watch by.

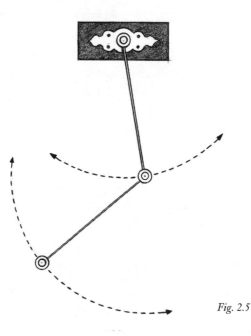

Fig. 2.5

Surprise. Chaos reigns. The path traced by the end of a double pendulum after it was released is an irregular squiggle (*Figure 2.6*). There is no simple, predictable pattern. And if we start from an even slightly different position, the chaotic squiggle will take on an entirely different look (*Figure 2.7*). Search for "double pendulum" on the Internet if you want to see animated simulations.

DRIPPING FAUCET. If we were to let a tiny bit of water drip out of the faucet of a sink, then surely we would find a regular rhythm to the beats of the drops.

Surprise. By adjusting the faucet slowly, we will find that at some settings we get regular dripping, but further adjustments of the faucet yield an unpredictable drip pattern in which the droplets fall

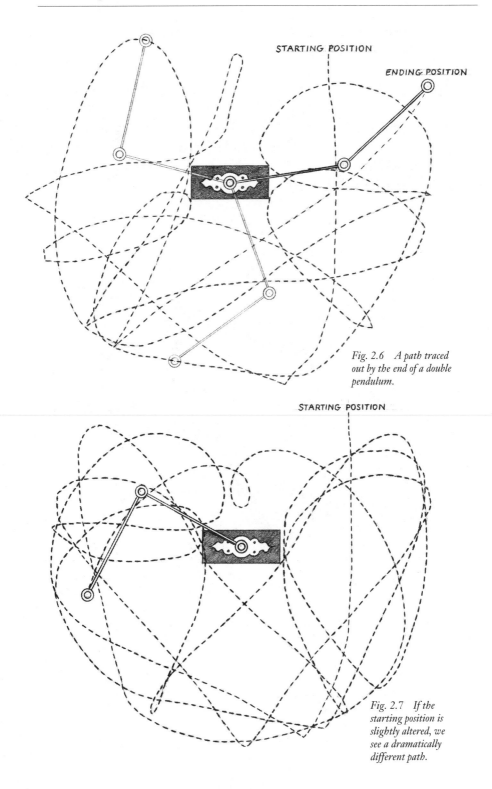

STARTING POSITION

ENDING POSITION

Fig. 2.6   A path traced out by the end of a double pendulum.

STARTING POSITION

Fig. 2.7   If the starting position is slightly altered, we see a dramatically different path.

at irregular, chaotic intervals. This simple dripping faucet confirms our sense that bathroom leaks lead to sheer chaos.

MAGNETS. Let's consider a hanging pendulum with a magnetic end that can swing freely in any direction and a base containing three mounted magnets that pull the pendulum toward them (*Figure 2.8*). Of course, when the pendulum gets closer to a magnet, that magnet's attraction is more pronounced. We might expect the tip of the pendulum to trace out a predictable regular pattern.

Fig. 2.8

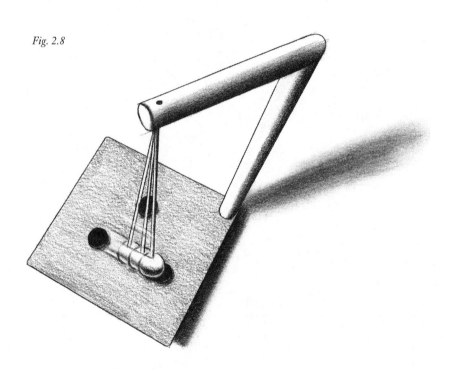

Surprise. We are again faced with chaos. The pattern traced is irregular, jerky, and unpredictable. For three different but close starting positions, the magnetic tip is attracted to a different magnet in each case, thus tracing out a very different pattern (*Figure 2.9*).

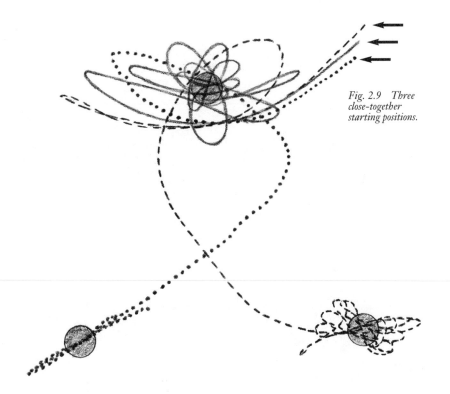

Fig. 2.9 Three close-together starting positions.

This simulation depicts the paths made by a magnetic pendulum for three slightly different starting positions. The tip of the pendulum ends up being attracted to a different magnet in each case.

BOUNCING BALL. If we drop a ball and let it bounce on the ground, we will witness a regular pattern of diminishing heights to the bounces (*Figure 2.10*). But let's consider a piston in a tube in which the piston is moving up and down at a regular rate. Suppose now we

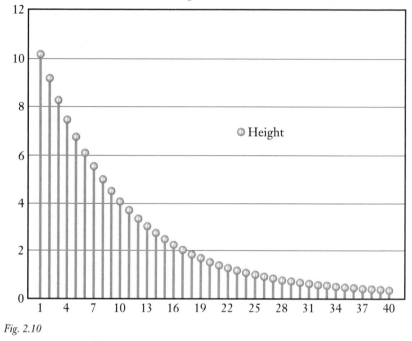

Fig. 2.10

drop a ball into the tube so it bounces up and down on the moving piston (*Figure 2.11*). Is there a rhythm to the heights reached by the bouncing ball?

Surprise. Chaos yet again: the pattern of the heights of the ball's bounces is chaotic (*Figure 2.12*). Of course, by now we can't really expect any reader to feel surprised at seeing chaos. And that is the point.

So mathematical chaos is indeed reflected in real life. These examples—pendulums, faucet drips, bouncing balls—are physical manifestations of the numerical chaos we saw earlier, and they are just a few illustrations of a pervasive theme. Before Lorentz's observations became known, physicists generally believed that the world described by classical mechanics was rather deterministic. But during the 1970s and 1980s, mathematicians and physicists began to explore the implications of the concept of chaos. It soon became clear that chaos—that is, sensitivity to initial conditions—was a common phenomenon both in the mathematical descriptions of the

Fig. 2.11

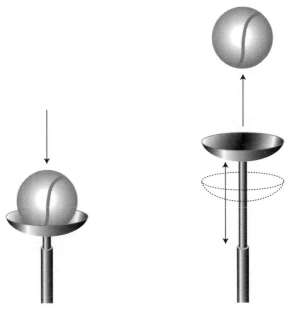

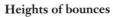

Heights of bounces

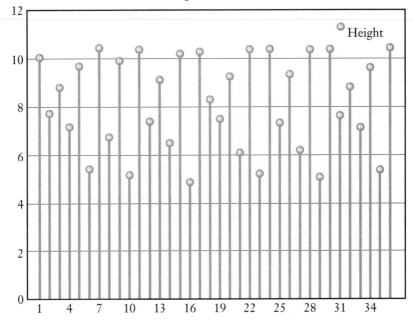

Fig. 2.12

world and in the world itself. Since even simple mathematical systems exhibit chaos and lead to unpredictability, physicists were forced to accept the idea that chaos was plausibly a fundamental feature of the reality of nature.

This insight has vast implications, even to the epistemology of science—that is, to our concept of the fundamental limits of scientific knowledge. Formerly, many physicists thought that quantum physics most profoundly demonstrated the uncertainty of scientific truth, and that perspective may be correct in some basic way. But perhaps chaos in classical mechanics limits our potential to predict the future of physical systems even more than quantum mechanics does. We will always have inaccuracies in measuring physical aspects of the world, and those inaccuracies will inevitably get magnified to produce dramatic differences in what we predict will happen in the not-too-distant future. Chaos suggests that there are severe limitations on our potential to predict the future behavior of even rather simple physical phenomena.

CONCLUDING CHAOS

When we try to understand the world, we sometimes attempt to develop reasonable mathematical models that allow us to fathom the future. We hope that the local conditions and influences that we see around us will allow us to see the world on its way—an emerging future determined by the circumstances of the present. But when we look at those models even in the abstract setting of pure mathematical logic, where nothing deviates from the truth and every calculation is unerring and precise, we still find that our view of future prospects is very foggy. The sensitivity to initial conditions in iterated or repeated mathematical processes leads to chaos. Since iterated systems are commonly used to describe behavior in many different fields, the implications of this study of chaos are far-reaching, affecting our understanding of weather, population, fluid dynamics, economics, the stock market, oscillating chemical reactions, electrical networks, and even heart rhythms and brain waves.

We cannot see the flickering flame of the distant future because we accumulate such a mass of tiny deviations along the way that the final destination is lost in the haze. Even mathematics, complete and precise, is subject to the perils of tiny variations in initial conditions, which, when multiplied and magnified by the tyrant of repeated application, end by leading us far astray. What consolation do we have? Only that the models can become slightly more perfect so that the inevitable deviations are pushed farther into the laps of generations to come—let them deal with the chaos!

CHAPTER 3

DIGESTING LIFE'S DATA

Statistical Surprises

There are three kinds of lies: lies, damned lies, and statistics.

—Benjamin Disraeli

Obviously . . . Not many Americans have one testicle and one ovary.

Surprise . . . In fact, the *average* American has one testicle and one ovary. Statistics is a subject of enormous utility that can lead us to great insights into trends and patterns. But blind application of statistical formulas can paint a misleading picture of our world and cause us to statistically drop the ball.

THE JOY OF DATA

What area allows us to discuss sex, drugs, and death openly and with such quantitative glee? Welcome to statistics. Much of our understanding of the world is based on statistical evidence or impressions. We develop expectations based on experience. Sometimes our intuition accurately reflects reality and, unfortunately, sometimes it doesn't.

Valid statistical inferences are important in making personal and societal decisions. But statistics is susceptible to misinterpretation. Some assertions based on data sound valid but, in fact, are accidental mistakes or deliberate lies. We can laugh at these pitfalls—unless, of course, we're the victims.

In this chapter we will explore instances of being duped by numerical shenanigans in the media, our schools, our friends, air travel, medical tests, and even spinning pennies. And what better place to start an exploration of statistical folly than within an arena famous for both mistakes and lies—U.S. presidential elections?

POLLING FOR PRESIDENTS

If you wanted well-digested literature in the early part of the twentieth century, then the *Literary Digest* was the publication for you. Its fame endures, however, owing to a statistical fiasco that it unwittingly perpetrated in 1936, one that will keep the *Digest* alive in statistics textbooks for generations to come. The setting was the 1936 U.S. presidential election; the *Literary Digest*'s mission, to predict the outcome. Most people living today did not vote in that election, but those who were around may remember that the two principal candidates were Franklin Delano Roosevelt, the incumbent, and Alfred Landon, the Republican opponent.

For each of the previous five presidential elections, the *Literary Digest* had correctly predicted the winner and had been within a few percentage points of getting the actual voting margin right. For the 1936 election, the *Digest* sent out millions of surveys to voters across the country. The evidence was clear and the prediction confidently proclaimed: Landon would easily win the White House. In fact, the periodical predicted that Alf would receive 57% of the popular vote and would win the Electoral College vote in a landslide, 370 to 161.

Most people will not recall studying President Alfred Landon in their United States history class, for the very simple reason that Landon didn't win. The *Literary Digest*'s predictions were correct in only one respect: the election *was* a landslide—but one that went the other way. Roosevelt was overwhelmingly reelected with 62% of the popular vote and won in the Electoral College by a remarkable vote of 523 to 8. How could the then-soon-to-be-ex-statisticians at the *Literary Digest* have been so utterly mistaken? Simple: They asked the wrong people for their opinions.

Lists of people to whom to send surveys were compiled by the *Literary Digest* from several sources, including subscription records for the *Digest*, car-registration records, and telephone records. Ten million surveys were sent and two million were returned. But 1936 fell in the middle of the Great Depression. Many households were cutting back on unnecessary expenses, and sadly (for the health of the *Literary Digest*), subscriptions to that esteemed publication may have been among the first casualties of the budget ax. Many families cut further still and did without cars and telephones. So the people on the survey list were not representative of the entire voting public. Furthermore, only the surveys that were voluntarily returned were counted. Who knows whether the people who return surveys are a cross-section of the public? In any case, the survey was wildly biased and the deductions made were grossly wrong.

GEORGE GALLUP. The *Literary Digest* fiasco had another interesting outcome—it catapulted a young statistician to enduring fame. George Gallup became aware of the *Literary Digest*'s survey and its methods. Suspecting that the survey would be faulty, he himself surveyed 50,000 people using a method that would later become a standard feature of good polling: pure randomness.

He chose his survey participants randomly from voter registration rolls and found that his survey presented a dramatically different picture. He not only predicted a strong Roosevelt victory, but also predicted what the *Literary Digest*'s prediction would be. So he announced, in advance, that the *Literary Digest*'s prediction would be wrong—and he was right. George Gallup saw the defect in gathering data from rich people (which led him to become rich himself). What he may not have foreseen is that "Gallup Poll" would become a household word.

The *Literary Digest* made the error of gathering its information from biased sources. Today, experts are more sensitive in their statistical studies, but they are still prone to misleading conclusions, so we must remain on our statistical guard. The *Literary Digest* fiasco is an example of poor statistical methods. No professional today would make that particular error—except intentionally. Suppose, for example, you wanted to author a convincing report supporting your view that almost all Americans drive dangerously. One way to collect data to support your stance is to simply make your observations in the parking lot of a defensive driving school—but be careful where you stand.

PERSONAL BIAS

Biased surveys from the political arena can be amusing and are important, but it might be worth a moment's thought to acknowledge that our personal impressions of the world are gathered largely from enormously biased sources: our friends, our neighbors, and the news.

While our friends are wonderful people, what they are not is a cross-section of humanity. Part of the wonderfulness of our friends is that they tend to agree with us. Of course, there are differences and there are exceptions, but all in all, the people we tend to be drawn to are people who have the wisdom and insight to see eye to eye with us on many issues. If we ask some of our friends and neighbors for their views and compare the proportions of opinions with the reported proportions in the news, we are often astounded. Haven't we all heard an interview on the news that made us think, "I've never met anyone silly enough to have that ridiculous opinion," only to learn later that the majority of people hold that same ridiculous point of view?

Newspapers, TV, radio, and other news outlets are all dramatically biased—and not in the way we might initially think. Of course, there are political, religious, or cultural biases that appear in media reports; however, the most egregious bias is one that all media outlets embrace: They are strongly skewed toward the *unusual*. When we watch the news, what we are really watching is a report on the strangest and rarest events of the day. We'll never see a newspaper headline that reads "PSYCHIC PREDICTS LOTTERY NUMBER THAT LOSES."

The other clear bias in the news is that it strongly favors the bad, which the public finds more sensational and exciting. Read the first section of the paper or watch the nightly news on television with an eye for good versus bad news. We will find that if we were to deduce the activities of our world from the news, we would conclude that almost everything happening around us is bad. Newspaper sales would plummet if the paper reported that yesterday the following people did *not* commit murder or rape, abuse their children, or defraud their investors. The bottom line is that we would rather hear the bad news than the good news; that is, bad news sells. The effect of the bad-news syndrome is that often we have a wildly inaccurate view of the world.

We are especially likely to hear bad news that seems "exotic" or outside of our everyday experience. As an illustration, we have heard for decades about terrorism in Israel, which remains an unfortunate reality. In 2002, for example, 238 people died in terrorist attacks in Israel. However, to put that number in perspective, we note that the danger of dying in an automobile accident in the United States in 2002 was three times the danger of dying in a terrorist incident in Israel in 2002. That is, the number of deaths per million people was three times as high for automobile accidents in the United States as for terrorist attacks in Israel, but we didn't hear much about the car accidents at home. We can't help but have a grossly inaccurate view of reality when the news we receive has to be, first and foremost, interesting—that is, unusual and bad.

In our own lives, having an accurate view of statistical reality is helpful so we can assess the risks we take. Every day we run risks. They come with the territory; we can't simply choose not to take them. Why? Because we run risks even when we decide to do nothing. In fact, as we all know, sitting around like potatoes is bad for our hearts and bad for our mental health. If we actually leave the comfort of our couch, we run different risks.

Let's say, for instance, that our driving skills are fine, but we have a relaxed view of the importance of obeying annoying traffic laws. The yellow light is to us what the red cape is to the bull—we charge right through. This bovine behavior is only mildly risky. Suppose we are likely to gore some poor unsuspecting driver only one in a thousand times. However, if we adopt this behavior once daily, then that 1-in-1,000 occurrence is statistically likely to happen to us roughly once every three years. So a driver who regularly accepts an apparently slight risk, one that results in an accident only 1 in 1,000 times, is really a bull in a china shop.

DREAM SCHOOL

Nothing is more important than the education of our children. A good education can turn lazy, video-obsessed, money-craving teenagers into the refined, successful, and socially responsible children of our dreams. What school will mold our malleable clay children into polished sculpted gold?

As parents we are not only loving but above all practical, so naturally we seek a school that will deliver the goods. Culture, refinement, and human compassion are one thing, but money in the bank is easier to measure. Why beat around the bush? We simply go for the cold, hard cash and let the cultural chips fall where they may.

For each school in the country, we compute the average net worth of its graduates. We find that in general the graduates of the "better" schools—the ones with reputations for excellence—draw higher salaries, but when we see the stats for one in particular, Lakeside, we halt our search and mail in our child's application immediately. After all, in a recent year, the average annual salary for a Lakeside School graduate was more than two million dollars! That is certainly the school for any child who may, at some later date, wish to support his or her loving parents.

Our hearts sink when we interview most of the living alumni and discover that not a single one of them has an income even close to the two-million-dollar mark. Our hopes for future financial bliss sink as we realize the error in our methodology. We looked at the *average* income of alumni. What we did not realize is that Bill Gates and Paul Allen attended Lakeside School. Their income that year was so vast that even if no other graduate earned a dime, the average for all alumni would still be over two million dollars. The average of earnings means almost nothing when these two multibillion-dollar outliers are thrown into the mix. Of course, Lakeside School is, in fact, an excellent school, and its graduates are far more successful than average; however, the $2,000,000 average annual salary is misleading.

RELIVING THE UPS AND DOWNS OF THE SAT

Every year, hundreds of thousands of high school juniors and seniors flock like sheep to the slaughter to the pre-college rite of passage known as the SAT. The SAT summarizes the worth of the first 17 years of a young person's life as a number between 400 and 1600. For several grueling hours on a Saturday morning, college-aspiring high school students wield their number No. 2 pencils in a desperate display of concentration, thought, and random guessing. Some have spent hundreds of their parents' hard-earned dollars on SAT vocabulary courses so that when they do poorly, they can exclaim, "My perspicacity at circumlocution evanesced," rather than "I forgot how to b.s."

But even the weakest student has the faint hope that luck alone will carry the day. If Ouija boards and divining rods work so well, why can't the silly pencil find the right oval to fill in on its own? Even students who don't know a proctologist's orifice from an aperture in the magma can harbor the faint hope that fate will lead the pencil point when the mind is a tabula rasa. If every student filled in the bubbles randomly, how would we expect the scores to look?

Let's assume that every question is multiple-choice with five possible answers. Statistically speaking, guessing randomly would result in getting 20% of the answers correct. Of course, the SAT people have thought of this, and that is why the scoring involves penalties for incorrect answers. But let's simply consider the simpler question of what percentage of questions the purely guessing test-taker is likely to get right.

On average, a random guesser would get 20% correct, since one out of the five choices is correct on each question. But some guessers, by random luck, would get more than 20% correct, and others would get less than 20% correct. If thousands of clueless students took the SAT employing the random-guessing strategy, we would expect to see the histogram of the number of questions correct look like a bell-shaped curve centered at 20% (*Figure 3.1*). Some students might get 23% correct. Some might guess correctly on only 18%. There is a remote chance that someone would get 70% correct by pure luck. In fact, there is a tiny, tiny chance that someone would guess every single question correctly. We should ask that lucky student to purchase our next lottery ticket.

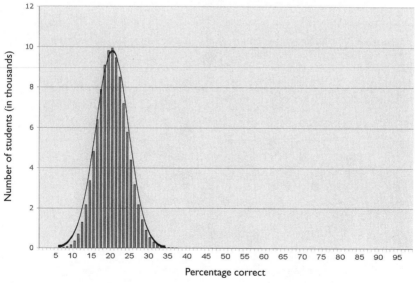

Fig. 3.1

The result of large numbers of students guessing answers on a multiple-choice test is an example of an approximation of a *normal distribution*. The effect of many small random factors influencing the global outcome produces a body of data that centers around one value and then tapers off on the sides. We can physically see this phenomenon in an apparatus known as a quincunx (an excellent SAT vocabulary word, by the way). A quincunx is a device made of a lattice of wooden pegs affixed to a board (*Figure 3.2*). A ball is dropped from the top center and bounces randomly off the pegs on its way down. Since at each striking of the ball with a peg, the ball is as likely to bounce off in one direction as the other, a dropped ball will, on average, end up under where it started. However, some of the balls will accidentally hit and bounce right more times than left. These balls will end up to the right of the center. Likewise, those that bounce left more times will end up left of center.

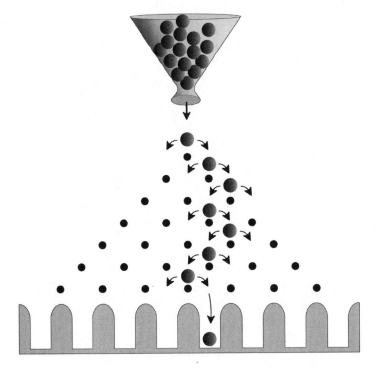

Fig. 3.2

Since a good number of bounces occur while the ball falls, there are many left-or-right random choices being made. Sure enough, if we drop a large number of balls down the quincunx, we find that the places where they land give a satisfying physical realization of the bell-shaped curve that we would expect from the logic of the situation (*Figure 3.3*).

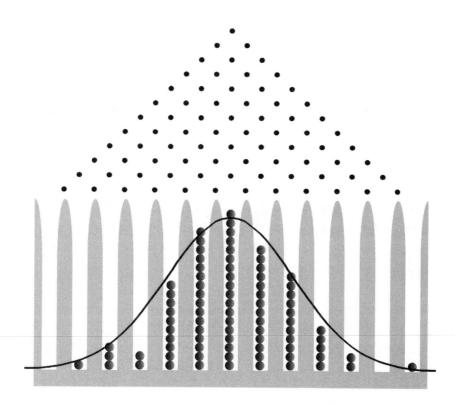

Fig. 3.3

A PENNY FOR YOUR STATISTICAL THOUGHTS

In our SAT scenario we saw that if we drop the ball (or in fact thousands of balls) and answer randomly, then a bell-shaped normal curve accurately depicts the distribution of the outcomes and confirms our hunches. In some instances, however, the data runs counter to our intuition. We illustrate this unexpected divergence with the lowly penny. Pennies don't buy much, but they can help with decisions. When we can't decide, we might well dig out a penny, flip it, and let fate enter the fold. In some statistical sense, we wish to learn which way Lincoln leans on the issue. That metaphor suggests a way to tilt the scales.

BALANCED THINKING. Instead of flipping that penny, we might consider a more delicate method of finding the balance of fate. On a table, let's carefully balance a penny on its thin edge. In fact, let's take 100 pennies and balance them all on their edges on that table. This exercise results in a highly unstable situation. We can stabilize the situation by simply slamming our hand on the table and letting the pennies fall where they may. Some of the pennies will lie face up; some will lie tails up. Of course, we expect roughly an equal number of heads and tails.

Surprise. If you perform the experiment, you will find that significantly more than half of the pennies land heads up. A balanced penny, surprisingly, is not equally likely to tip over on either side; its not-quite-symmetrical construction gives it a slight tendency to lean more toward landing heads up.

THE SPIN CYCLE. Balancing pennies is hard work, so we might prefer a more dynamic method of allowing the penny to put a spin on our future. In fact, we could try literally spinning a penny on a table. The spinning penny will eventually grow weary, slow down, and finally fall in dizzied relief. As before, we repeat this experiment 100 times to gain a statistical sense of the outcome. Again, we would expect about an equal number of the pennies to fall heads up as tails up.

Surprise. This time the pennies' penchant is for tails. That is, the asymmetrical characteristics of pennies cause spun pennies to land tails up more frequently than heads up. So pennies balanced and pennies spun each provide examples where outcomes we expect to be equal turn out to be quite different.

The practical result of these surprising examples with pennies is that if we play our cards right, we will rarely have to pay for dinner with friends again. When the check arrives and it's time to decide who pays, just say to your friend, "Let's let Honest Abe decide. Choose heads or tails." If your friend says, "Heads," then you say, "You know, flipping a penny gets so boring; let's spin a hundred pennies and see how they land." A majority will probably land tails up, and you will win and not have to pay for dinner. If your friend says, "Tails," then you respond, "Instead of flipping a penny, which is so passé, let's balance a hundred pennies on edge, slam our hand down on the table, and see how they land." A majority will probably land heads up, and you will win and not have to pay for dinner. Either way, fate is on your side.

Perhaps the true lesson here is that our intuition may suggest that events will fall one way, but reality might have another plan in mind. We need to acknowledge life's data, let the chips fall where they may, and keep an open mind. Sometimes we need to re-educate our intuition when it comes to guessing likelihoods.

REUNIONS

Another counterintuitive conundrum makes a surprise appearance in a reunion scenario. Suppose we attend our twenty-fifth college reunion after not seeing our classmates since graduation. The first thing we observe is that everyone else is old. Unfortunately, they are looking at us with the same sympathetic horror. The second interesting thing that happens is that we overhear the following conversation between two reunited acquaintances:

BOB: Joe told me you have two children.
BETTY: That's right.
BOB: I thought he also mentioned that your older child is a boy.
BETTY: That's right again. I must confess I didn't know Joe was such a yenta!

At the moment of Betty's confession, the pig-in-a-blanket that she was attempting to swallow begins to put up a valiant fight. She starts to choke and turn colors that clash with her blue dress. As we watch Bob perform the Heimlich maneuver, a position he always dreamed about being in with Betty, we cannot help but wonder: What is the probability that Betty has both a boy and a girl?

As we ponder that puzzle, we overhear another conversation:

VICTOR: So Joe told me you have two children.

VIOLET: That's right.

VICTOR: I thought he also mentioned that one of your children is
         actually named Victor.

VIOLET: That's right. Victor is a typical teenage boy—and damn,
         doesn't Joe ever mind his own business?

At this moment in Violet's colloquy, she meets the same fate as
her former roommate Betty—she too finds her hors d'oeuvre too
tough to swallow. As she vainly attempts to clear her throat, we are
reminded of the line, "Violets are blue." Again we cannot help but
wonder: What is the probability that Violet has both a boy and a
girl?

EXPECTED EQUALITY. Upon first blush of blue, it appears that
these two slightly different conversations present the same situation.
In both cases, we know that one of the two children is a boy, and the
gender of the other child remains the unknowable secret of our col-
lapsed classmate. Surely, then, the likelihood that Betty or Violet has
a boy and a girl is 50-50 or $\frac{1}{2}$, since one child is definitely a boy and
the other child is equally likely to be a boy or a girl.

Surprise. In fact, these two nearly identical dialogues yield two
quite different likelihood scenarios. To detect the difference, we
parse the repartee line by line. At the outset of each conversation, we
learn that our old classmate has two children. At that point we do
not know anything about the genders of her children, so there are
four equally likely possibilities:

1. The older child is a girl and the younger is also a girl.
2. The older child is a girl and the younger is a boy.
3. The older child is a boy and the younger is also a boy.
4. The older child is a boy and the younger is a girl.

In her conversation with Bob, Betty reveals that the *older* child is
a boy. Given this new information, we know that possibilities #1
(older girl, younger girl) and #2 (older girl, younger boy) cannot be

true. While she's choking on the pig in her throat, we gain no further information, so we realize that there are two equally likely possibilities for Betty's children, namely the two remaining possibilities above: #3 (older boy, younger boy) and #4 (older boy, younger girl). Since there are two equally likely possibilities, one of which is that Betty has a boy and a girl, we correctly conclude that the likelihood of Betty's having a boy and a girl is $\frac{1}{2}$ (one out of two possibilities).

In the dialogue between Violet and Victor, we again learn early on that Violet has two children, so again we have the same four equally likely possibilities in mind. This time, however, we learn that Violet has at least one son. Armed with this new knowledge, we can rule out possibility #1 (girl, girl). All three of the other possibilities are still viable, since the boy that we know about could be either the older child or the younger child. Among those three possibilities, two of the three consist of both a boy and a girl; thus we conclude that the probability that Violet has both a boy and a girl is $\frac{2}{3}$ (two out of three possibilities).

This reunion story offers a striking illustration in which two situations seem identical in their essential features (of two children, one is a boy and and we don't know the gender of the other), and yet the two cases actually differ dramatically. If we were buying presents for the children, understanding the differences in the two situations would encourage us to buy a football helmet and a tutu for Violet's children rather than a pair of athletic supporters. Or, come to think of it, we might just go with slide rules for both families.

Beyond the antics, when we strip away the joking and the choking, the reunion scenarios illustrate some strategies of analysis that are effective methods of clear thinking: clarify the issue and isolate the essential ingredients. Thus while we cannot know for sure the gender breakdown of either pair of children, we can apply numerical measures that allow us to make some intelligent guesses.

We close this chapter with two serious real-world illustrations of the subtleties of probability. Here again, a careful analysis of the data forces us in each case to see the issue in a new light.

AIR SAFETY

About the safest place we can be is 30,000 feet above sea level flying in a U.S. commercial airplane. It is so rare for a commercial airliner to crash that every instance stays in our collective memories for years. Of course, all crashes are tragedies, so every responsible citizen must wonder whether air travel could be made even safer. Let's explore the possibility of improving airliner safety.

First, how safe in fact are airplanes? We'll quantify this notion by considering the number of people who have died in commercial airline accidents and by determining, on average, how many passenger-miles were flown per death. (A passenger-mile is one mile flown by one passenger—so a 2,000-mile flight with 100 people on board would count as 200,000 passenger-miles.) The record is pretty amazing. American commercial airlines fly approximately 1.7 billion passenger-miles per day. Over the last decade or so, on average there have been approximately 183 deaths per year in commercial airline accidents, which is about one death every two days or one death per 3.4 billion passenger-miles. Of course, the reason that airline accidents are so spectacularly horrific is that people don't die one at a time, spaced evenly throughout the year. Instead, there is a tragic accident in which hundreds of people are killed, and then a year passes before another tragic accident occurs. A once-a-year mass death is more memorable than a dribble of one death every two days.

Let's try to put the rate of one death for every 3.4 billion passenger-miles in perspective. Suppose a particular individual flew 1,000 miles each day. On average, it would take that person 3.4 million days of flying—that is, 3.4 billion divided by 1,000—before he or she was killed in an airline accident. That many days comes to roughly 10,000 years. So that person has a long wait and a huge mound of tiny bags of pretzels ahead before he or she is likely to be part of a commercial airline disaster.

Airlines are phenomenally safe; however, we could make them safer still. Better maintenance, better training and screening of pilots, better air-traffic-control equipment and policies, and better movies on board would all make airlines safer. (Well, okay, the last one would just make air travel more bearable.) In any case, perhaps with diligence we could make air travel ten times as safe as it is today. That is, we could make the average number of deaths per year plummet from 183 to just 18. Such a policy would save 165 lives per year. One of them might be ours, so let's contemplate this great-sounding idea.

UNINTENDED CONSEQUENCES. The trouble with silver clouds is that they often have a sewer lining. In this case, let's consider the implications of our wonderful airline-safety proposal. Airline safety has a price, and that price is cash. Airplane ticket prices will rise. And when ticket prices rise, some people who now fly will stop flying and a certain number of those non-flyers will decide to drive to their destinations instead. Uh-oh . . .

Let's expand our scope of thinking to include those additional people who will be tooling along our highways instead of tossing back tiny pretzels at 30,000 feet. For argument's sake, suppose that 10% of the people who previously flew would now choose to drive because of the higher cost or greater hassle of air travel. So in our new system, an extra 170 million passenger-miles will be *driven* rather than flown each day. What effects will that have? To compute the consequences of this increase in driving, we first note that car deaths are about 34 times more frequent per passenger-mile than deaths during air travel. So these 170 million extra passenger-miles per day have a real consequence. In fact, since there is an automobile death every 100,000,000 (one hundred million) passenger-miles, we would expect, on average, to see about 1.7 more automobile deaths per day than before we enacted our air-safety policies. Thus, over the course of a year there would be a total of approximately 620 additional automobile deaths. Of course, there would be 165 fewer airline deaths, but the net consequence of our airline-safety idea is that we would lose roughly 455 more lives per year than before our safety measures. Whoops.

WHY MAKING AIRLINES LESS SAFE MIGHT SAVE LIVES. If air travel could be made significantly less expensive, then many people who now drive would fly. The net effect would be that lives would be saved. One way to make air travel less expensive would be to *relax* aspects of air travel that contribute to safety—the quality of the airplanes, maintenance standards, rules about how close planes can fly to one another and how frequently they can take off and land at airports, and so on. Making air travel cheaper and slightly less safe might well result in fewer people dying overall, since fewer people would be involved in the far more dangerous act of driving. The moral of this paradoxical tale is that we should always explore unintended consequences—especially when we are considering public policy decisions.

HIV/AIDS—UNIVERSAL TESTING, UNIVERSAL TRAUMA

AIDS (acquired immune deficiency syndrome) is the disease that results from being infected with HIV (human immunodeficiency virus). AIDS has been a worldwide epidemic for the last two decades; it is the leading cause of death among young people. Given the seriousness of the disease and the fact that people with HIV may not exhibit symptoms of AIDS for many years, testing people for HIV would provide us with important information for combating the disease at both the personal and the societal level. Perhaps we should enact a policy of testing every man, woman, and child in the United States. Before pushing such a policy, however, let's size up the statistics.

The sobering figures are these. In 2001, approximately 42 million people worldwide had HIV/AIDS, and about 3.1 million people died as a result of AIDS. In the United States, roughly 500,000 people are living with the virus.

Tests for HIV are quite reliable. The ELISA blood-screening test, for example, tests positive for 95% of people who actually have the disease, and it tests negative for 99% of people who do not have the disease. That seems very good and suggests that universal testing might reliably tell us who is infected, so that those people could be warned to avoid behavior that would further spread the disease. Given all this data, should we undertake universal testing for HIV/AIDS?

Let us consider the consequences of instituting such a program. Suppose a random person takes the test and then receives the dreaded news that the test was positive. The question is: What is the probability that a person who has a positive test result is actually HIV-positive? The reliability of the test seems to answer this question. Recall that when an uninfected person takes the test, the test returns a (false) positive result only 1% of the time and returns a (true) negative result 99% of the time. This statistic seems to state that a person who gets a positive test result is 99% certain to be HIV-positive.

But if we crunch the numbers, we draw a surprisingly different conclusion. Here are some facts:

1. The population of the United States is approximately 280,000,000.
2. The number of people in the United States who are HIV-positive is about 500,000.
3. The number of people in the United States who do not have the virus is 279,500,000 (that is, 280,000,000 minus 500,000).
4. Of the 500,000 people infected with the virus, the test will detect the disease in 95% of the cases—that is, for 475,000 people.
5. Of the 279,500,000 people who do not have the virus, the test will report falsely positive in 1% of the cases—that is, for 2,795,000 people.
6. The total number of people receiving a positive test result is 3,270,000 (that is, 475,000 plus 2,795,000).
7. Of the 3,270,000 who receive a positive test result, only 475,000 actually have the disease.

Therefore, a person who receives a positive test result has only a 475,000/3,270,000 chance of actually having the disease. Let's express that ratio as a percentage. Getting a positive test result means that the person still has *less than a 15% chance* of actually having the disease! Thus we arrive at a peculiar paradox: The test is 99% accurate, but when a person receives a positive test result, that person has less than a 15% chance of actually being infected. How can this be?

One way to reconcile this conundrum is to note that there are approximately 475,000 people in the United States who have HIV and get a positive test result, while roughly 3,300,000 people in all receive positive test results. So only about 1 in 7 positive test results can be correct.

Notice that the positive test result does substantially alter our view of the *probability* of having the disease. Since only 1 in 500 people in the United States has HIV, then a random person, with no other information given, would be assumed to have only a 1-in-500, that is, one-fifth of one percent, likelihood of having the disease. However, after the person receives a positive test result, our estimate for the probability of his or her being infected with the disease has risen to 15%—a substantial increase.

Of course, in practice we don't live with the uncertainty; when a test comes back positive, we take more refined tests so as to find out for sure. But before that is accomplished, we worry. So one consequence of universal AIDS testing would be that millions of people would receive frightening false positives and would need to have further testing.

One conclusion we can draw from this scenario is that clear thinking is absolutely essential when we are making policy decisions. The results may well include significant and harmful unintended consequences.

STATISTICS ON BALANCE

Statistics can help us understand the world. It is a powerful and effective tool for placing economic, social welfare, sports, and health issues into perspective. It molds data into digestible morsels and shows us a measured way to look at situations that have either random or unknown features. But we must use common sense when applying statistics or other tools that draw on our experience of the world to shape data into meaningful conclusions. Carefully thinking through deductions that follow from some statistical arguments might help us avoid the perils of statistical tomfoolery. While the numbers may not lie, the manner in which they are offered to us could allow them to fib just a little.

In the next part of the book, we move from seeing numbers as a statistical measure to focusing on them in their own right. As we'll soon discover, numbers have individual personalities that we can embrace and secrets that we can attempt to lure out of the shadows.

PART II

EMBRACING FIGURES

Sensing Secrecy, Magnificent Magnitudes,
and Nature's Numbers

Next we journey from our uncertain and chaotic everyday world to the exacting and regular world of numbers. Our daily interactions with numbers are usually fleeting as we use them to quickly quantify a notion or measure an object. Here we slow down and take a moment to get to know the personalities of numbers. After all, numbers figure prominently in our lives. Some are secret, such as our Swiss bank account numbers. Some are big, such as the age of the universe in seconds. Some come from nature, such as the spiral counts within the center of a daisy. Here we will explore numbers—their secrets, their sizes, and how they appear in nature.

Secrets are among the curiosities of life. We long to know what is withheld, and we long to keep our own confidential communications with our accountants, lawyers, and banks out of the curious and prying eyes of our enemies and the IRS. But secrets are also the property of the universe and create a thread that weaves through the ages. Ancient mysteries involving numbers tantalize us and will keep our minds engaged for centuries to come.

When it comes to numbers, size *does* matter. When we think of numbers we can't help but be drawn to the question of size. How many stars are in the night sky? How much does a million dollars weigh? How thick would a piece of paper folded fifty times be? (Thicker than you'd think.) At a million dollars an hour, how long would it take to pay off the national debt? How many handshakes connect us to Thomas Jefferson? Even rough estimation helps frame our conception of the universe—its age, its size, its contents—and of our everyday lives. Orders-of-magnitude approximations help us put our life experience in perspective.

When we look closely at objects in nature, such as the spirals on pineapples, we find numerical intrigue. The numbers in nature are not arbitrary and random. They exhibit a pattern and a structure that nature embraced long before humankind entered onto the scene. Exploring mathematical ideas that stem from the numbers suggested by nature reveals patterns in the abstract that lead us to artistic and aesthetic insights as well as to a deeper appreciation for the beauty within nature.

Being attuned to the numbers that arise around us is a powerful way of seeing our world with more nuance and detail. Numbers have a way of forcing us to see structure and meaning in our surroundings—structure that could remain hidden without the magnifying lens of quantitative focus.

CHAPTER 4

SECRETS HELD, SECRETS REVEALED

Cryptography Decrypted

Why are wise few, fools numerous in excesse?
'Cause, wanting number, they are numberlesse.—Augusta Lovelace

Obviously . . . In the dark cloak-and-dagger days of the Cold War, undercover agents would gather top-secret information, encode that intelligence using top-secret codes, and then smuggle the microfilmed message to their governments via clandestine drop boxes. Keeping the intelligence, the encoding procedure, and the coded messages secret was paramount for those deep-cover spies.

Surprise . . . Today, spies can announce their encoding methods to the entire world and broadcast their top-secret encoded messages in newspapers and on the Internet. Paradoxically, this public encryption scheme, in which everyone has access to both the encoding process and the encoded messages, is even more secure than (if not as seductive as) 007's romantic method of passing bits of microfiche to attractive spylets via a French kiss. Sorry, James.

Spies and our universe have something in common—both hold secrets that we don't know. Some of the most delectable delights of mental life involve mysteries and secrets. When we look at an encrypted message, we know it holds a secret that we could read if only we had the key to unlock it; and when we investigate unknown questions in mathematics, we sense that the universe herself is holding secrets that we might be able to glimpse if only we looked at them the right way.

The enticement of secrets comes from knowing a little but not the whole tamale. Tantalizing crumbs of partial disclosure lead us through a forest of deep, dark secrets. Each crumb, each bit of information, gives us a hint about the truth up ahead. In some cases we can measure how close we are to that truth; in other cases we cannot really guess whether the whole truth lies just an inch away or whether no human being will ever know it.

In our exploration of secrets held and secrets revealed, we turn first to the world of espionage and the Internet, where secrets are routinely held and transmitted with the aid of unusual secret codes—secret codes that are made public.

SENDING SECRETS

When we think of secrets and how they are handled, it isn't long before we think about codes. Secret codes are not employed only by spies; in fact, all of us non-spies use secret codes every day without realizing it. The coding scheme we will now explore is one of the methods used in practice by computers whenever we perform an on-line banking transaction, use an ATM, make an Internet stock trade, or type our credit card number on a Web site. The key to this modern means of communicating sensitive information revolves around numbers. Who would have thought that the simple process of multiplying two numbers together would lie at the heart of Internet commerce and the confidential transmission of secret information?

As long as there have been enemies, there have been secret codes. Julius Caesar, who had a good number of enemies (including Brutus), devised his own coding scheme, now known as a *Caesar*

Cipher. People who have played the Cryptoquote word game in newspapers know this scheme—one letter is substituted for another. For example, "HELLO" might be coded as "QLAAJ" (here "H" is replaced by "Q"; "E" by "L"; "L" by "A"; and "O" by "J"). There are three significant drawbacks to the Caesar Cipher. The first is that it might, as Caesar himself discovered, indirectly lead to our assassination. The other two problems are less fatal but still serious. One is that the code is very easy to break; the other is that the code would be completely compromised if the codebook fell into the wrong hands.

The fact that the code is easy to break is demonstrated daily as millions of newspaper readers break the short coded messages just for the fun of it. Breaking the code is really equivalent to stealing the codebook. Obviously, if the codebook fell into the hands of Mark Antony, he would be able to decode any coded message merely by reversing the coding process. (Wherever he sees a "Q," he replaces it with an "H," and so on.) The ease of breakability cries out for a more complicated coding scheme. However, the problem of a stolen codebook appears impossible to resolve—it seems clear that if someone knows the procedure to *encode* a message, then performing that procedure in reverse would allow one to *decode* encrypted messages.

Surprise. A counterintuitive coding scheme exists in which the knowledge of exactly how a message was encoded still does not allow one to decode it. Only the designated receiver, who has an added piece of information, can decode the message. In fact, we could make the encoding process completely public—that is, announce the process to the entire world—without compromising its effectiveness. How is such a paradoxical coding scheme possible? The answer is as easy, or as hard, as factoring a number.

The heart of the issue is: How is it possible that the knowledge of the method for encoding messages does not automatically give away the decoding process? The answer to this fundamental question is not only the key to understanding how such a counterintuitive coding scheme is possible, but the key to unlocking the specific details of the scheme itself.

As always, let's start off simply. Suppose someone named Simon announces that he is thinking of the number 15. So far we would not rate this news as "breaking." If simple Simon then reveals that the number 15 could be written as a product of two smaller numbers, we would probably not alert CNN; it is easy to see that $15 = 3 \times 5$. What's the point? Well, suppose that Simon says that he's now thinking about a bigger number that is also the product of two smaller numbers, and that new number is 33. Again, we can easily see that $33 = 3 \times 11$. Let's now view the two smaller numbers—the factors of the first number—as Simon's secret. Of course, with 15 and with 33, it's not much of a secret, since we quickly realize that 15 is 3 times 5 and that 33 is 3 times 11.

Let's suppose now that Simon starts thinking of larger numbers that are each the product of two smaller numbers. If Simon says 247, then those two factors are not as easy to find as before. With a bit of effort we would discover that $247 = 13 \times 19$. If Simon says 1219, then it would take even more effort for us to realize that $1219 = 23 \times 53$. And if he announced 1,308,119, then we'd be forced to pull out our calculators to finally find that $1{,}308{,}119 = 661 \times 1979$. Even most calculators would be of no help if Simon announced 17,158,904,089, although a computer could crack it and discover that $17{,}158{,}904{,}089 = 104{,}729 \times 163{,}841$.

But if Simon announced

13284718321501923414108776182374182348172341874 31
83257612381823471823748123748123764812735172635 47
26345837457812376172465123491234577128345761872 36
45187234561827345671263498172347612374619278364 59
12783456473827364657283712873645198273569128475 69
18237417263549182734198273498127365491287345127 36
45182734659182374199162358164182374971265348172 63
45987162398745619823756918273456182736482371872 34
812634731,

then the fastest computer in the world would take longer than the history of the universe to find its two factors by any known method.

This scenario illustrates an interesting dichotomy between theoretical knowledge and practical knowledge. We are given a number—for example, this number with 401 digits—and we are informed that Simon knows that it is the product of two smaller numbers. In *theory*, we could find those factors by trial and error, but that procedure would take a ridiculously long time. From a practical point of view, then, even though we know that Simon's number *can* be factored, its actual factorization remains unknowable.

PUBLIC KEY CODES

That simple and perhaps trivial-sounding observation is the key to *public key cryptography*. Suppose that Simon wishes the world to know that he is open to receive coded messages. He announces a huge number, such as the previous 401-digit number, that neither person nor computer can factor, and he announces the arithmetical instructions on how to use that number to convert a message into an encoded message. Decoding an encoded message sent to Simon involves arithmetical steps that require the secret factors of his large public number, as we will see later. Since no one other than Simon knows those two factors, no one can decode the encoded messages sent to Simon except Simon himself. Thus, encoded messages do not have to be secret—they can be published in newspapers or posted on the Internet, and only Simon will be able to make any sense out of the gibberish.

So what are the coding and decoding procedures, and how does knowledge of the factors allow the decoding? After giving an overview, we'll offer a complete example using small numbers so that you can see for yourself how it works.

Let's say we wish to send the message "HELLO" to Simon. We begin by converting that message to numbers. In our system, each letter is converted to the two-digit number corresponding to its position in the alphabet—that is "A" is 01, "B" is 02, and so on, up through "Z" = 26. So our message, "HELLO," is converted to the number 0,805,121,216. Note that we keep the zeroes; we just string together all the two-digit numbers, in this case 08, 05, 12, and so on. Next we raise that number to a very large power (announced publicly by Simon). Finally, we take the enormous answer and divide it by Simon's publicly announced 401-digit number, leaving a remainder. That *remainder* is the encoded message that we send to Simon. Of course, in practice, computers do all the calculations for us. Simon receives the encoded message (actually, just a number), raises it to a secret power that involves the two secret factors, and then finds the remainder when that huge number is divided by his 401-digit number. Surprisingly, the answer will be the number 0,805,121,216, which Simon can easily convert back into letters that spell out our "HELLO."

In summary, to encode a number, we raise that number to a publicly given power and then compute the remainder after dividing by a second publicly announced number. The remainder found is the encoded number. To decode that encoded number, Simon raises it to another power (one that only he knows) and again computes the remainder after again dividing by the second public number. That remainder will, amazingly, be the original number, thereby decoding the message.

Next we offer an explicit example involving small numbers to illustrate exactly how the factors of the huge number come into play in the coding scheme. Some readers will be content with the overview just given. We invite those sensible readers to pass over the following section. Other readers might enjoy seeing a hint of the numerical process in action. For those hardy souls, this section's for you.

PLAYING THE NUMBERS

Suppose Simon publicly announces the numbers 7 and 33 and publicly explains how to use them to encode a message. He tells us that to encode a message to him, we should first convert the letters into a number, raise that number to the 7th power, divide the result by 33, and find the remainder. The remainder is the encoded message that we send to Simon. So, for example, suppose we wish to send the secret message "E":

(1) We change "E" to 05.
(2) We raise 5 to the 7th power: $5^7 = 78,125$.
(3) We divide by 33 and find the remainder: $78,125 \div 33 = 2,367$ with a remainder of 14.
(4) So, 14 is our encoded message.

Now Simon decodes the encrypted message by raising it to the 3rd power, dividing the result by 33, finding the remainder, and converting that number back to a letter. For reasons explained below, Simon (and only Simon) knows that raising the encoded message to the 3rd power is the key to deciphering it. Let's watch Simon go at it with the encoded message 14:

(1) He raises 14 to the 3rd power: $14^3 = 2,744$.
(2) He divides by 33 and finds the remainder: $2,744 \div 33 = 83$ with a remainder of 5.
(3) He converts the number 5 to "E."
(4) So, Simon has decoded the secret message "E."

Try this encoding and decoding for yourself, using a calculator. Start with any number less than 33. You'll discover that raising that number to the 7th power and finding the remainder when you divide by 33 and then raising that remainder to the 3rd power and finding the remainder when you divide by 33 will give back the same number you started with. Mysterious magic? Nope—just marvelous mathematics.

So how do the factors of 33 enter into the process? We notice first that $33 = 3 \times 11$ (and that 3 and 11 cannot be factored further—they are *prime* numbers). Next we multiply one less than the first factor times one less than the second factor and add one: $(3 - 1) \times (11 - 1) + 1 = 21$. Since that result can be factored, $21 = 7 \times 3$, then one of those factors, 7, can be used as the encoding power, with the other factor, 3, as the decoding power. This simple example is a very special case of a much more general procedure that continues to hold with much larger values.

In real life, the steps in the coding and decoding process involve numbers with hundreds of digits, and the message tends to be a lot longer than "E" or "HELLO." The sizes of the numbers that Simon announces—one for the encoding power, one for dividing by to yield a remainder—depend on the current factoring abilities of computers. Computers have finite capabilities, and thus at any given time there exist numbers so big that they cannot be factored by any computers of the day. To create a truly unbreakable code, we need only select two (prime) numbers, both so enormous that their product far exceeds any computer's factoring ability. But how can computers multiply those obscenely large numbers together if they cannot factor the result? The answer is that multiplying numbers or raising numbers to enormous powers and finding remainders is computationally easy even when the numbers have hundreds or thousands of digits, while factoring numbers with several hundred digits is computationally very difficult or, in practice although not in theory, impossible.

HOW MUCH IS YOUR INFORMATION WORTH?

Two important issues that naturally arise actually hang together: How large should Simon's number be? How valuable is the information that Simon receives? The more valuable the information, the larger the number should be. The smaller the number, the easier it is to break the code. For example, the coding method associated with 33 is broken instantly, since anyone can see that $33 = 3 \times 11$. It requires more time and effort, however, to break the code associated with 17,158,904,089. A computer would certainly be needed, and the information would have to be sufficiently valuable to make it worth the effort to crack that number open. On the other hand, if what we're encoding with that 11-digit number is national security information, then the modest effort required to determine that $17,158,904,089 = 104,729 \times 163,841$, and thereby break the code, is certainly worth the enormous payoff to some national enemy. Thus, for important information, we would require a substantially larger number to ensure security.

The study of the value of information is an extremely important one in our technological times; hence the value of public key encryption schemes is high. As we'll discover in the next chapter, to mathematicians, a number such as 400,000,000 is considered a tiny value relative to all numbers. However, there is an easy way to make that value become large in the eyes of everyone, including mathematicians—place a dollar sign in front of it: $400,000,000. Roughly ten years ago, Security Dynamics paid four hundred million dollars to purchase an encryption company that developed this public key encryption method. Not a bad haul for some mathematicians who put a few arithmetically simple steps together in a profoundly insightful manner.

IS BREAKING UP REALLY HARD TO DO?

If we could factor Simon's number, then we could break his code. Finding some easy way of factoring numbers would compromise the code. Today no one knows if there exists some practical method of quickly factoring enormous numbers. But there might be some clever method of factoring that would allow the most anemic computer to factor our 401-digit number in seconds. If some computer thief finds such a method, we will all experience havoc of a scope that only the Internet can create.

In a slightly different direction, is there perhaps a devilishly cunning method of breaking the encryption scheme *without* factoring the number? No one knows the answer to this extremely important question. However, once again the fact that we don't know the answer means that we cannot break the code. So until we can overcome our current state of ignorance, this encryption scheme is safely shrouded in secrecy. Public key encryption is a prime example of exploiting our ignorance to our advantage.

THE UNIVERSE'S SECRETS

Schools often present mathematics in dull and distorted ways. In particular, many students are left with the impression that all of mathematics is understood and all mathematical questions have been answered. Many people believe that there is no mathematics beyond the lofty heights of calculus. But in reality, while there is a considerable body of mathematics beyond calculus that mathematicians have come to understand, almost all of mathematics remains outside our grasp. Despite centuries of effort, the universe continues to hide answers to most mathematical issues.

In fact, it is a little embarrassing to acknowledge some of the simple-sounding questions that no mathematician can yet answer. Even within the most basic building blocks of numbers, the universe has kept answers to easy questions secret and hidden from our view. We close this chapter on secrets by exploring several mysteries that nature refuses to reveal.

NUMBERS FOR THEIR OWN SAKE. For thousands of years, people have loved numbers and found patterns and structures among them. The allure of numbers is not limited to or driven by a desire to change the world in any practical way. When we observe how numbers are connected to one another, we are seeing the inner workings of a fundamental concept. These observations often come from people who simply love numbers. The numbers speak to them.

But sometimes the numbers do not speak loudly enough for us to hear. In those cases we hear tantalizing whispers of possibilities without the satisfaction of completion. The theory of numbers is full of elementary questions that remain unanswered to this day.

Recall that a *prime number* is a whole number greater than 1 that cannot be written as the product of smaller whole numbers. The first half-dozen or so primes are 2, 3, 5, 7, 11, 13, and 17. The number 6 is not a prime, since it can be expressed as 2×3. Every whole number bigger than 1 is either a prime number or the product of prime numbers. The prime numbers are the building blocks—or the elements, so to speak—for the whole numbers; they are the indivisible units from which all other numbers are made. As such, primes have been one of the most studied ideas in human history. But there are questions about prime numbers that have baffled mathematicians for hundreds of years and remain unanswered. Let's explore two of them.

THE GOLDBACH CONJECTURE. *Every even number greater than 2 is the sum of two primes.*

This conjecture, formulated by Christian Goldbach on June 7, 1742, remains unresolved to this very day—that is, no one knows whether it is true or false. If we consider some examples, we see that for small even numbers it seems to be true: $4 = 2 + 2$; $6 = 3 + 3$; $8 = 3 + 5$; $10 = 3 + 7$; $12 = 5 + 7$; $14 = 3 + 11$; and $16 = 5 + 11$.... In fact, using computers and abstract arguments, mathematicians have verified that every even number up to roughly 10^{17} is the sum of two primes. While that is a huge list of numbers, there are still infinitely many more to verify. To illustrate how far we are from resolving this issue, we note that in 1939 it was shown that every even number can be expressed as a sum of no more than 300,000 prime numbers! Obviously, we have a way to go to bring 300,000 down to 2.

Suppose someone someday proves the Goldbach Conjecture or produces a counterexample, that is, produces an even number that is provably not the sum of two primes. Would it make any practical difference? Our first guess is, probably not. What possible difference could that answer make? Often mathematical discoveries seem totally disconnected from practicality. But as history has shown us, mathematical issues that appear to be completely divorced from our everyday lives today, may hold the key to our everyday lives tomorrow. The ancient idea of primes, together with some abstract insights that were discovered 350 years ago, provided the basis for public key cryptography. Perhaps a resolution to the Goldbach Conjecture will lead to a windfall of $400 million in a hundred years. (Unfortunately, in a hundred years, $400 million might have the buying power of only $24.99.)

For two years, actually, the solution was worth $1,000,000: A foundation offered a prize of a million bucks if someone could verify the Goldbach Conjecture between March 20, 2000, and March 20, 2002. Our guess is that the foundation wasn't very concerned about the possibility of having to cough up a giant check—and indeed the foundation didn't have to.

When we look at the prime numbers in order, we are led to the next famous mystery regarding the primes. Occasionally we find two odd prime numbers that are as close as numerically possible—that is, two odd primes flanking the even (non-prime) number between them: 5 and 7; 11 and 13; 17 and 19; 29 and 31; 41 and 43. We call these pairs of primes *twin primes*.

TWIN PRIME CONJECTURE. There are infinitely many twin primes.

This conjecture has been the subject of thought by mathematicians for centuries, but we don't know yet whether there are infinitely many twin primes or whether at some point they run out. The largest pair of twin primes currently known have 51,090 digits each. They are:

$$33218925 \times 2^{169690} - 1 \quad \text{and} \quad 33218925 \times 2^{169690} + 1.$$

Is it useless to find out whether the Twin Prime Conjecture is true? Who knows—perhaps it is the key to some cosmic safe that we have yet to encounter.

HEADING DOWN A ROAD TOWARD 1. Here's a silly game. Think of any whole number and then perform the following easy steps:

1. If the number is even, divide it by 2.
2. If the number is odd, multiply it by 3 and then add 1.
3. Repeat the process unless the answer is 1.

Let's try it out. Suppose we start with 19. Since it's odd, we multiply by 3 and add 1, which yields 58. Since 58 is even, we divide it by 2 to get 29. Since 29 is odd, we multiply by 3 and add 1, which is 88. Since 88 is even, we divide by 2 to get 44. We again divide by 2 to get 22 and again divide by 2 to get 11. Since 11 is odd, we multiply it by 3 and add 1 to arrive at 34. We now divide by 2 to produce 17. Since it's odd, we multiply by 3 and add 1 to yield 52. Since 52 is even, we divide it by 2 to get 26. We again divide by 2 to get 13. Since 13 is odd, we multiply by 3 and add 1 to arrive at 40. We now divide by 2 to produce 20. Divide by 2 again to get 10. Divide by 2 again to get 5. Multiply by 3 and add 1 to get 16. Dividing by 2 gives 8; dividing by 2 again gives 4; dividing by 2 again gives 2; and finally, dividing by 2 gives 1. Now we're at 1, so we stop.

If we try the process with a few more random numbers, we will find each time that we eventually get to 1. In fact, using high-speed computers and some theoretical work, people with far too much time on their hands have shown that for every number up to around 317×10^{15}, the procedure eventually leads to 1.

Mystery. If we do this process of repeatedly multiplying by 3 and adding 1 if the number is odd and dividing by 2 if the number is even (known as the *3x + 1 procedure*), then will we always end up at 1?

No one knows. Not many people care. But a few strange creatures find this question (and others like it) so intriguing that they stay up late pondering possibilities and methodically beating their heads against a brick wall until the wall or their heads slowly begin to give way. If they answer the question, will it make any difference? Who knows? But we do know that our instinctive desire to wonder about the world of numbers has paid enormous practical dividends in the past—abstract ideas about primes and factoring unexpectedly led to public key cryptography and security in Internet commerce. Somehow human curiosity about numbers from ancient times to the present seems to be in synchronicity with the universe.

CHAPTER 5

SIZING UP NUMBERS

How Many? How Big? How Quick?

Wherever there is number, there is beauty.—Proclus

Obviously . . . Suppose we fold an ordinary piece of paper in half. Now we fold it again, and again, and so on. While in practice the paper can be folded in half repeatedly only about seven times, let's imagine folding it roughly fifty times. When we're done, how thick is that folded sheet? An inch? A foot? A yard?

Surprise . . . The folded paper you created in your mind would be a tower rising up over 100,000,000 miles (yes, that's one hundred million miles) and would extend past the sun. Not bad for about fifty folds . . .

NUMBERS

When most of us think of math, we first think of numbers. And when we think of numbers, we first think of counting. While at first blush we may not view the act of enumeration as profound, it is certainly something we can always count on. (Sorry.) Contrary to our initial impression, the process of counting has a potency far beyond its seeming simplicity. Few abstract ideas in human history have given us greater power to see the world in a nuanced way. Discovering *how many* is like looking into a microscope—suddenly we see the world in far greater detail. In this chapter we'll discover that the simple act of counting holds the key to many surprising outcomes.

There are three types of people: those who count precisely and those who don't. Of course, in reality, what counts as counting depends on the scenario—sometimes we feel compelled to count carefully, and at other times a rough estimate will suffice. For example, we tend to count precisely when we leave the airport and want to ensure that we have all our luggage and all our children in tow. However, we forgo the exact count and consider "orders of magnitude" when we contemplate the redness of the national debt or the greenness of Bill Gates's wealth—that is, we care how many zeroes there are at the end of the number (say, ten billion versus a hundred billion), but we don't care whether the number preceding the zeroes is a 1 or a 2 (ten billion versus twenty billion). As it turns out, both types of counting—exact or order-of-magnitude estimating—can better inform our view of the world.

One well-known example of a quantity that numbers measure either precisely or as an estimate is age. For a period that lasted more than 40 years, the celebrated comedian Jack Benny estimated his age to be 39—a classic comedy bit, but not the most accurate estimate in history. A few years before the birth of Jack Benny was the birth of Earth; and in the seventeenth century Bishop James Ussher computed the exact day on which the Earth was born: October 23, 4004 B.C. He obtained this startling accuracy using astronomy and the Bible to journey back through all the recorded *begat*-ings until he arrived at Adam and Eve. However, Ussher's vision of a rather youthful Earth (roughly 6,000 years young) did not survive the test of time. In the nineteenth century Charles Lyell, a geologist, used scientific evidence to more accurately "card" Earth and determine that it was many millions of years old. Thanks to Dr. Lyell, Earth was then eligible for the senior discount at the movies.

Today, scientists believe the Earth to be a few billion years old, and their modern methods are unlikely to be radically wrong—although Bishop Ussher may have felt the same way about *his* methods in the seventeenth century. Our concept of the world is fundamentally altered when we view Earth as millions or billions of years old rather than a few thousand years young. For instance, assuming an advanced age for Earth gives natural phenomena—the gradual ones, such as erosion, and the rare ones, such as earthquakes and volcanoes—enough time to produce their dramatic effects: the creation of canyons, mountains, continents. So an order-of-magnitude concept of the age of the Earth changes our perspective regarding what influences cause Earth's frequent face-lifts.

NATURAL NUMBERS

Natural numbers are the numerical foundation on which our quantitative sense of both mathematics and our world rests. The *natural numbers* are the counting numbers—that is, 1, 2, 3, 4, 5, and so on forever. Of course, the stream of natural numbers flows on endlessly.

There are two types of natural numbers: the small ones and the large ones. We know and love the small numbers because we feel comfortable with them and encounter them day in and day out. The other numbers appear foreign and formidable.

It is interesting to consider the fact that most natural numbers are *really* big and thus have absolutely no true meaning to us. In actuality, almost all natural numbers have never been encountered by any human being, dead or alive. Here we unveil a virgin number, untouched by human eyes until her appearance in this book. Let us present our digital debutante:

50588773485839972782674565498949117196545 6687
65212321654545497646412141916564654464654 5641
49848949498198949876313213465462198498198 1004
18098787818009098741596512026125450040885 0998
77001128588907542123500205569868189602005 4567
13389654911631166400688464013145504788877 2325
99524838798255259159871258059885254155859 8852
22699632268987046544646545641498489494981 9894
98763132134654621984981981004180987878180 0909
87415965120261254500409825525915987125805 9885
25415585988522269963226898704654464654563 7209
60071.

Actually this numerical nymphet is simply a random string of 500 digits. The probability that someone wishing to write a 500-digit number would randomly produce that exact value is $1/10^{500}$. How rare is that? Suppose that every person on Earth, each one of the 6.4 billion of us, is entered in a worldwide raffle. Once a week, one winner is selected at random from that enormous pool of humanity. It is more likely that you would win this worldwide raffle *every single week for an entire year* than that you would randomly produce a 500-digit number that exactly matches the one displayed above.

Of course, there are infinitely many numbers, so there are infinitely many of them that no human being has ever seen; but some readers might wonder if someone had ever *counted* up to this particular 500-digit value. Isn't it possible that some Egyptian scholar, perhaps in the third dynasty, worked her way up to that exact number? Well, no. Let's see why.

The ancient Egyptians were amazing. They built pyramids that seem impossibly large and complex, given their technology, and even today, in the '80s retro dance clubs, people are still walking like them. Still, no member of this talented group could have reached this number. Suppose we moved that Egyptian scholar even further back in time—say, to the moment of the Big Bang, roughly 13.7 billion years ago—and armed her with a supercomputer whose counting ability was 1 trillion numbers per second. If that supercomputer counted away at that lightning speed since the birth of the universe and never stopped, how close would it have come to our 500-digit number by now?

It's easy to estimate how high our Egyptian supercomputer would have counted: We simply multiply the 13.7 billion years by 365 days per year by 24 hours per day by 60 minutes per hour by 60 seconds per minute by 1 trillion counts per second. Multiplying these quantities together gives us a number with a mere 29 digits. That number is so minuscule compared with our 500-digit number that it couldn't be viewed as even a good start. Isn't it amazing that the entire age of the universe in trillionths of a second can be measured with a number less than 30 digits long? A 500-digit number, for all practical purposes, doesn't mean anything. It's just too big.

NUMERICAL NAMES

The numbers we talk about, of course, have names. But actually, most natural numbers have yet to be named. The numbers that do have names are mere specks of dust when compared with most natural numbers. Perhaps we should take a lesson from those companies that, for $50, will name a far-off star after a loved one. Thus we might consider starting a company that, for a modest fee, would name a far-off number after a loved one, and record the name in some meaningless registry. Of course, our company would be in business long after those others run out of stars. In fact, no matter how many numbers we name for our number-loving clientele, we have made *no* progress in naming all the numbers—infinitely many of them remain unnamed. Given these thoughts, you might already be eyeing the 500-digit number made famous in this book and wish to name it in honor of a dear relative; perhaps we can call it the Aunt Edna Number. (Please remit a money order.)

PUTTING A FACE TO A NUMBER

A 500-digit number has no meaning to us. But our modern world, particularly with our omnipresent computers, forces us to confront numbers in the millions, billions, and even trillions. Understanding the differences among these enormous quantities sometimes has real consequences. To develop an intuition about the distinctions between millions and billions and trillions, let's explore some scenarios where they arise in life.

THOUSANDS. A lofty example of one thousand is the approximate number of stars visible to the naked eye in the night sky. In fact, in the 17th century, Tycho Brahe and Johannes Kepler recorded 1,005 stars (although they never thought of selling them). Of course, your star count may vary. The number of hours a student spends in class during a college education is one or two thousand; that number is also approximately the number of hours we sleep in a year. Coincidence? We think not . . .

MILLIONS. To make even more unwieldy quantities meaningful to us, we can embark upon some thought experiments. To make millions meaningful, let's think of some collections that are measured in millions. The populations of Los Angeles and New York are measured in millions. A high-resolution digital picture has a few megapixels. *Mega* means roughly a million; perhaps that is why we refer to show-biz personalities who are worth millions of dollars as mega-stars.

But let's try to get a more concrete sense of a million by pondering the weighty question "How much would a million dollars weigh?" That is, suppose some impish individual offered us a million dollars in one-dollar bills, but with the proviso that we alone must carry them away all at once. Is this easy money? Exactly how much does a million dollars weigh?

Of course, if we had a stack of a thousand one-dollar bills lying around, then we could just weigh it and multiply by a thousand. Since most of us tend not to have that many singles in our possession, let's consider a more commonly available collection of paper items: a ream of copier paper, which weighs roughly 4 pounds and contains exactly 500 sheets. Any counterfeiter we might ask would tell us that a sheet of copier paper could fit about five bills. So roughly 5 × 500 or 2,500 dollar bills would weigh about the same as a ream of paper. A million is 2,500 × 400. So we estimate that a million one-dollar bills weighs approximately 4 × 400 or 1,600 pounds! Even Governor Arnold Schwarzenegger would have a tough time carrying away a million dollars in ones—an unfortunate realization, since the great state of California could sure use the cash.

BILLIONS. The most notable collection measured in billions is all of us—that is, all of humanity—who number about 6.4 billion. And the age of the universe since the Big Bang is, as we noted above, 13.7 billion years. A gigabyte means roughly a billion bytes, so our hard drives hold another example of billions.

But these examples of billions do not allow us to truly wrap our minds around such magnificently enormous quantities. To really picture billions, let's think about Bill Gates. A few years ago, Bill Gates enjoyed a year in which his personal wealth increased by twenty billion dollars. We can put that awe-inspiring income in perspective by imagining the following scenario.

One day during the year that Mr. Gates is making twenty billion dollars, he's toiling away in his cubicle and he sees a hundred-dollar bill lying on the floor. Would he be better off going off the clock for the moment required to pick it up, or should he ignore the bill and just keep earning his salary?

Since it might take Mr. Gates a second or two to reach down and pick up the $100 bill, we need to calculate his wage per second. If we assume that he works a 40-hour week for 50 weeks, then in the course of the year he works 2,000 hours. Each hour has 3,600 seconds, so he works 2,000 × 3,600 or 7,200,000 seconds. Since he is earning $20,000,000,000, his "secondly" salary is $20,000,000,000/ 7,200,000—that is, roughly $2,800 per second.

Therefore, every $\frac{1}{28}$th of a second, he earns $100. Olympic sprinters with extraordinarily fast reflexes require roughly a tenth of a second to react to the starting gun, which is more than twice as long as it takes Mr. Gates to earn $100. So not only should Mr.

Gates not clock out to pick up the $100 bill, he shouldn't even stop to look at it. Billions are pretty big.

SARDINES. We constantly read that the world is overpopulated. But perhaps the only problem is where the population is living. Suppose we gathered up all of humanity for an annual convention downtown. Could they all fit in the conference center? Could they even fit in town?

Surprise. The entire population of the planet could fit in one tall townhouse. No, we're kidding. But in fact all 6.4 billion people *could* fit in one cubic mile.

Let's do the math. A mile is 5,280 feet. So a cubic mile is approximately 150 billion cubic feet (5,280 feet × 5,280 feet × 5,280 feet). How much space would each of the 6.4 billion people on Earth have? Well, 150 billion cubic feet divided by 6.4 billion people is roughly 23 cubic feet per person—plenty of room. While that one-cubic-mile motel might not have the amenities of The Plaza, we could all squeeze in.

TRILLIONS. Trillions don't arise too often in our daily lives. However, one dismaying example of trillions is the national debt of the United States, which in January 2004 was clocked at exactly seven trillion dollars, that is, $7,000,000,000,000. Now *that* is a big number. How can we wrap our minds around a number this enormous? One method is to imagine paying it off.

Suppose we became serious about debt reduction and a fiscally fastidious legislator fought hard for a piece of legislation that would reduce the debt by a million dollars *per hour*. That's some serious belt tightening, but at least we could see paying off that debt down the road. How long would it take us to pay off the debt at that rapid rate? Well, obviously, it would take $7 trillion (7,000,000,000,000) divided by 1 million (1,000,000) hours, that is, 7 million (7,000,000) hours. There are 365 days each year and 24 hours each day, so there are about 365 × 24 = 8,760 hours each year. So dividing the 7,000,000 hours required to pay off the debt by the 8,760 hours there are in each year, we see that at a million dollars an hour, it will take us approximately 800 years to pay off that debt. We don't need any Munchkin economist to inform us that we'll be following that *red* brick road for quite some time to come.

QUADRILLIONS. Our first step toward a quadrillion (1,000,000, 000,000,000) is a single fold in a sheet of paper. We then fold it in half again, and again, and repeat and repeat (*Figure 5.1*). If we could fold that piece of paper in half just 50 times—which, in practice, is utterly impossible—then that theoretical experiment would produce a stack of paper that is amazingly tall, and the number of its layers, a quadrillion, would dwarf the national debt. Let's see why.

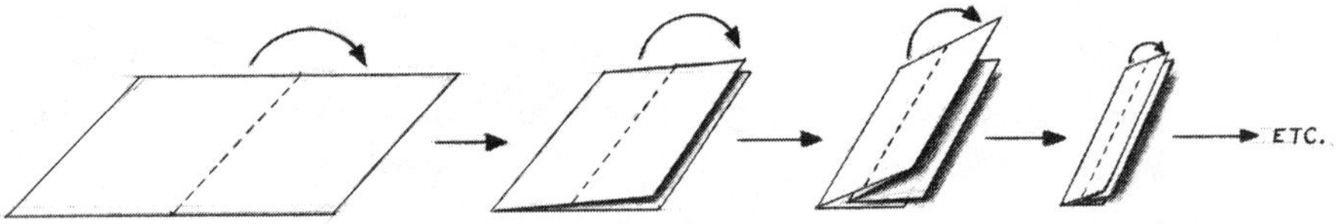

Fig. 5.1

When we fold the paper the first time, the folded sheet is two layers thick. When we fold it a second time, the folded sheet is four layers thick. Every time we fold the paper, the folded sheet has twice as many layers of thickness as before. After the first few folds, the number of layers begins to get really large:

> 3 folds—8 layers
> 4 folds—16 layers
> 5 folds—32 layers
> 6 folds—64 layers
> 7 folds—128 layers
> 8 folds—256 layers
> 9 folds—512 layers

After 10 folds, we'd have 1,024 layers. So if we could fold a sheet of paper 10 times, which we can't, the folded paper would be 1,024 sheets thick, which is approximately the thickness of two reams of paper—roughly 4 inches.

As we continue folding, doubling the thickness and number of layers with each fold, we notice that our tower is getting pretty tall pretty fast (*Figure 5.2*). After 20 folds, we have over a million layers with a total thickness that exceeds the length of a football field. After 30 folds, we have a billion layers that together are 64 miles thick. With 40 folds we reach one trillion layers of paper. After 42 folds, we are well past the moon. At 50 folds we have reached one quadrillion layers, and with one more fold the paper would be about 128 million miles thick, which is greater than the distance from the Earth to the sun!

Fig. 5.2

Folds	Layers (approx.)	Thickness (approx.)
10	1,000 *(one thousand)*	4 inches
15		11 feet
20	1,000,000 *(one million)*	350 feet
25		2 miles
30	1,000,000,000 *(one billion)*	64 miles
35		2,000 miles
40	1,000,000,000,000 *(one trillion)*	64,000 miles
45		2,000,000 miles
50	1,000,000,000,000,000 *(one quadrillion)*	64,000,000 miles
51		128,000,000 miles *[past the sun]*

Repeated doubling is an explosive process. With the paper-folding example we've come to the counterintuitive realization that within remarkably few steps we can generate enormous numbers. Next we will discover a counterintuitive reality in the opposite direction—namely, that among enormous quantities we find that any two are connected to each other within surprisingly few steps.

SIX DEGREES OF SEPARATION

Thinking numerically can bring us closer to our favorite heroes and heroines. Have you ever wished you could shake hands with Franklin D. Roosevelt, Margaret Thatcher, Marilyn Monroe, or Elvis? Although we won't meet every famous person we'd like to meet—especially the ones who died before we were born—we're closer to fame than we might initially guess. You are almost certainly fewer than six handshakes away from FDR or Thatcher or Marilyn or Elvis. That is, you have shaken the hand of someone (one shake away from you) who has shaken the hand of someone (two shakes from you) who has shaken the hand of someone (three shakes) who has shaken the hand of someone (four) who has shaken the hand of someone (five) who has shaken the hand of Elvis (six). You are so close to Elvis that you can probably begin to feel your hips gyrate as you read this very sentence. The world is definitely a smaller place than we might imagine.

How can it be true that we're all so closely connected to one another? Well, let's start with a famous person such as FDR. During his lifetime in the public eye, he shook hands with an enormous number of politically active people, including all the members of Congress, all the governors of all the states, military leaders, and prominent players in governments here and around the world, in addition to thousands of ordinary citizens. Those people in turn shook hands with other politicians, who in turn shook hands with activists and religious leaders. These people shook hands with their parishioners and indirectly with their neighbors. By this time nearly everyone is included. After they've all shaken hands with their cousins and in-laws, essentially everyone has been indirectly shaken by FDR.

You can estimate your own degree of handshake separation from anyone else. Suppose you want to estimate your shaking distance from the actor Kevin Bacon. You might begin by thinking of the most famous person with whom you have ever shaken hands. Or perhaps you know someone who has met a famous person. Famous actors and famous people in general either have shaken hands with each other or are within one shake of each other. So it is likely that it requires about only three shakes to get to a famous person and another three to get back down to any other person. It might take a few more shakes to reach a person who lives in a remote region in China.

We can even connect ourselves to our founding fathers with not too many additional handshakes. For example, let's consider Thomas Jefferson. He died on July 4, 1826, on the fiftieth anniversary of the signing of the Declaration of Independence. He almost certainly shook hands with several people who lived well into the twentieth century. Among those, someone in her eighties or nineties may have shaken the hand of a tiny baby who is still alive today. So it may well be that you are within six or seven handshakes of Thomas Jefferson, and certainly you are within eight or nine. In fact, some people alive today could conceivably have shaken the hand of someone who shook Jefferson's hand. The world is truly a global village. Whom would you like to meet? He or she is just a few handshakes away.

On the less pleasant side, this observation implies that we are only seven or eight handshakes away from the most repulsive or germ-infested human beings on the planet. Let's hope some of those intermediate shakers washed their hands before the chain reached us. Carrying this discussion slightly to the more serious side, the fact that we are all so closely connected is also the reason why communicable diseases are potentially so devastating. Suppose there were some disease that was lethal, symptom-free for some period, and transmissible by breathing or shaking hands. Then that disease would very soon infect nearly every person on the planet. It is sobering to realize that if HIV were transmitted by air, we'd all be infected by now.

SITTING ON OUR DECK

Leaving behind our crowded planet with its networks of people, let's sit alone for a while and play with cards. Suppose we place a very large stack of playing cards on a table and carefully cantilever them so that the top cards extend farther and farther over the edge of the table (*Figure 5.3*). Of course, as with most card games, we are not allowed to use any glue or other method for affixing the cards together. Given an unlimited supply of cards, how far can that stack extend off the edge of the table? Specifically,

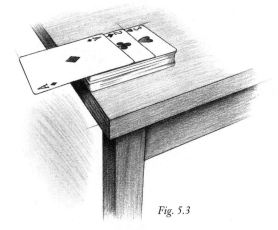

Fig. 5.3

1. Can we slide the cards so that the top card is completely beyond the table's edge (*Figure 5.4*)?

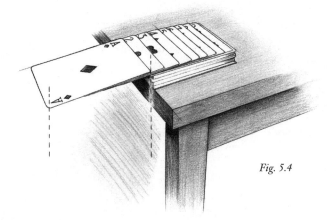

Fig. 5.4

2. Can we arrange the cards so that the top card is more than two
   feet beyond the end of the table (*Figure 5.5*)?

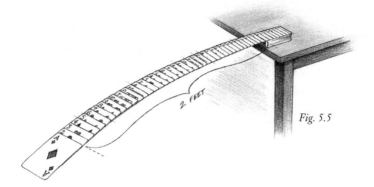

Fig. 5.5

3. Can we arrange the cards, again without the use of glue, so that
   the top card is more than a mile beyond the end of the table and
   so that we could actually sit on that top card without collapsing
   the leaning pile (*Figure 5.6*)?

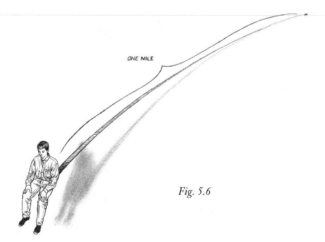

Fig. 5.6

Surprise. The answers are yes, yes, and yes! In fact, given enough
cards, we could theoretically stack them so that they extend as far
beyond the edge of the table as we wish. Furthermore, a pre-
assigned weight, no matter how heavy, could be placed on the top

card (the one farthest from the table) without the unglued deck of cards toppling over (*Figure 5.7*). This utterly counterintuitive fact is at once unbelievable and wonderful. Of course, we are operating in a hypothetical world where we can imagine a stack of a trillion cards or more, and we can imagine that that unrealistically huge stack of cards behaves with the same laws of physics as a real stack of 52 cards does. No real table could support such a stack of cards, but all the same we can describe what would happen if the table were unrealistically strong and the cards were unreasonably numerous. How is this feat possible? And how many cards would we need?

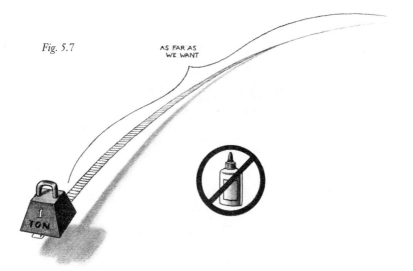

Fig. 5.7

AS FAR AS
WE WANT

The explanation that follows contains some numerical computations that might not appeal to everyone. We urge all readers who are uninterested in the technical details to either skim the prose or simply consider the images and move on. The unbelievable fact that we can fan out playing cards without glue so that they extend far beyond the edge of a table and then sit on the farthest card without the cards toppling over is something that we can all enjoy—even those of us who don't want to know the details of why it is possible.

LET'S TAKE IT FROM THE TOP. Instead of focusing on the bottom of the stack and moving up and out, we consider the top of the stack and work our way down. The key idea is that in order for the top card not to fall, its center of gravity must rest upon some point of the card that is directly beneath it. The *center of gravity* of an object is that point upon which the object would perfectly balance if placed on a pin. If the center of gravity of the top card is not over the lower card, then the top card will fall (*Figure 5.8*).

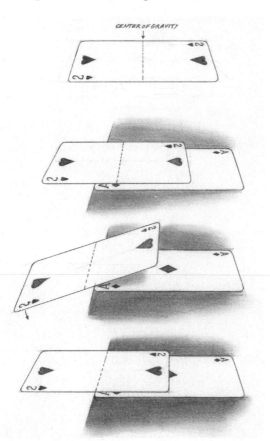

Fig. 5.8 If the center of gravity of the top card is over the lower card, then the top card stays in place. If the center of gravity is not over the lower card, then the card falls. If the center of gravity is at the edge of the lower card, then the top card balances perfectly and stays in place.

It is an easy matter to determine how far out the top card can sit relative to the card upon which it is perched. If more than half of it is hanging over the lower card, then of course it will fall over. So let's try to cantilever the cards as much as possible and see how the center of gravity of the cards moves with each subsequent (lower) card.

Let's assume that each card has a length equal to 2 units. Then the center of gravity of the top card is halfway along the length—1 unit from its edge (*Figure 5.9*). If we position the next card beneath it to

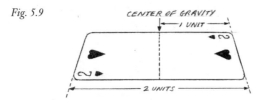

Fig. 5.9

have its left edge directly under that center-of-gravity point of the first card (*Figure 5.10*), then the center of gravity of the two cards together will be $\frac{1}{2}$ unit to the left of the center of the second card, since the second card's center of gravity will be 1 unit beyond the center of gravity of the first card. In fact, the center of gravity of the two cards together is actually equal to the average of the centers of gravity of the two individual cards.

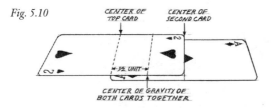

Fig. 5.10

We now place a third card underneath so that its left edge is exactly beneath the center of gravity of the top two cards (*Figure 5.11*). To determine the center of gravity of all three cards, we calculate as follows: We know that the center of gravity of the top two cards is precisely at the left edge of the third card and the center of gravity of the third card is, of course, 1 unit from its own left edge.

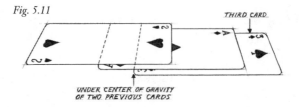

Fig. 5.11

So we have two card weights at the left edge of the third card and one card weight 1 unit to the right (*Figure 5.12*). Therefore, measur-

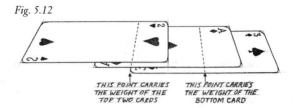

Fig. 5.12

THIS POINT CARRIES THE WEIGHT OF THE TOP TWO CARDS

THIS POINT CARRIES THE WEIGHT OF THE BOTTOM CARD

ing from the left edge of the third card, we average and find that the center of gravity will be $(2 \times 0 + 1)/3$ units to the right of the left edge of the third card (*Figure 5.13*).

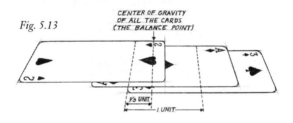

Fig. 5.13

CENTER OF GRAVITY OF ALL THE CARDS (THE BALANCE POINT)

⅓ UNIT

1 UNIT

Let's consider the placement of the fourth card from the top. Suppose we position it so that the top three cards will just barely balance on its left edge. That is, the center of gravity of the top three cards will be exactly above the left edge of the fourth card. Then the new center of gravity of the four cards will be located $(3 \times 0 + 1)/4$ to the right of the left edge of the fourth card (*Figure 5.14*).

Slowly a pattern is appearing. It is becoming clear that once we have positioned, say, 100 cards, we can put the 101st card under the

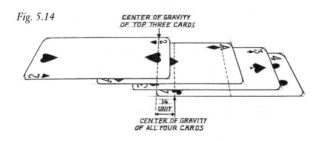

Fig. 5.14

CENTER OF GRAVITY OF TOP THREE CARDS

¼ UNIT

CENTER OF GRAVITY OF ALL FOUR CARDS

others so that its left edge is exactly beneath their center of gravity, and the top 100 cards will not fall over. The new center of gravity of the 101 cards will be $(100 \times 0 + 1)/101$ units to the right of the left edge of the 101st card.

As long as we continue to place the center of gravity of the entire stack of cards above the table, the stack will not fall. So if we want to know how far a stack of four cards can be cantilevered over the table, we would need to know how far the center of gravity of those four cards could be from the left edge of the top card (*Figure 5.15*). We see that the answer is:

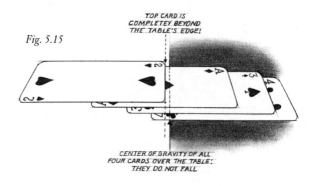

Fig. 5.15

TOP CARD IS COMPLETEY BEYOND THE TABLE'S EDGE!

CENTER OF GRAVITY OF ALL FOUR CARDS OVER THE TABLE: THEY DO NOT FALL

$$1 + \frac{1}{2} + \frac{1}{3} + \frac{1}{4} = \frac{25}{12} > 2$$

Since the length of a card is 2, this computation shows that the left edge of the top card is more than 2 units beyond the edge of the table and consequently that the top card is entirely beyond the table's top. Find four cards and try it! With some care, you can actually create this amazing sight.

Let's consider the other questions. Can the cards be stacked so that the top card is more than 2 feet from the edge of the table or more than 10 feet or more than a mile? The answer is yes in each case, because if we use enough cards, we can get the center of gravity of the entire stack to be as far to the right of the top card as we wish.

Suppose we wanted to create an unglued stack of cards with the top card more than a mile beyond the edge of the table. We'd need a lot of cards, but here is how we could theoretically make such a stack. It is easiest to show in terms of card units, so first let's express a mile in terms of card lengths. If one card is 2 units, as we discussed before, then one mile, which is 5,280 feet or about 60,000 inches, would be about 40,000 of our card units (assuming the cards are about 3 inches long).

We wish to show that we could balance the cards so that the top card is a mile—that is, approximately 40,000 card units—beyond the table's edge. With four cards we found that the distance between the left edge of the top card and the table's edge is the sum of the reciprocals of the first four natural numbers. This pattern continues. So if we had 100 cards, then the distance from the edge of the table to the left edge of the top card would be

$$1 + \frac{1}{2} + \frac{1}{3} + \frac{1}{4} + \frac{1}{5} + \ldots + \frac{1}{99} + \frac{1}{100},$$

which is equal to approximately 5.2 card units. As an aside, we note that with two regular decks of playing cards, we could get the outer edge of the top card more than two and a half card lengths beyond the edge of the table!

Our mission is now clear: Find the number of cards that would be required for the sum of all the reciprocals up to that number to exceed 40,000. To make the complicated summation a bit more manageable, we group the fractions into collections that each add up to at least $\frac{1}{2}$. If we found 80,000 $\frac{1}{2}$'s, then we'd have the total sum exceed the 40,000 level, as desired. Let's see how this grouping process looks.

$$1 + \frac{1}{2} + \left[\frac{1}{3} + \frac{1}{4}\right] + \left[\frac{1}{5} + \frac{1}{6} + \frac{1}{7} + \frac{1}{8}\right] +$$

$$\left[\frac{1}{9} + \frac{1}{10} + \frac{1}{11} + \ldots + \frac{1}{15} + \frac{1}{16}\right] + \left[\frac{1}{17} + \ldots + \frac{1}{32}\right] + \ldots$$

We claim that the sum of the fractions within each pair of brackets exceeds $\frac{1}{2}$. Let's see why this claim is valid for the first few pairs of brackets. Starting with the first:

$$\left[\frac{1}{3} + \frac{1}{4}\right] > \left[\frac{1}{4} + \frac{1}{4}\right] = 2 \times \frac{1}{4} = \frac{1}{2}.$$

This calculation is simply the observation that since two $\frac{1}{4}$'s equal $\frac{1}{2}$, then if we add together $\frac{1}{4}$ and something bigger than $\frac{1}{4}$, we'll get more than $\frac{1}{2}$. The same reasoning can be applied to the next set of brackets. We notice that:

$$\left[\frac{1}{5} + \frac{1}{6} + \frac{1}{7} + \frac{1}{8}\right] > \left[\frac{1}{8} + \frac{1}{8} + \frac{1}{8} + \frac{1}{8}\right] = 4 \times \frac{1}{8} = \frac{1}{2}.$$

So the four terms from $\frac{1}{5}$ through $\frac{1}{8}$ add up to more than $\frac{1}{2}$ again. Similarly, the eight terms from $\frac{1}{9}$ through $\frac{1}{16}$ sum to more than $\frac{1}{2}$. The next 16 terms (from $\frac{1}{17}$ through $\frac{1}{32}$) add up to more than $\frac{1}{2}$; the next 32 terms (from $\frac{1}{33}$ through $\frac{1}{64}$) add up to more than $\frac{1}{2}$; and so on forever.

Therefore, if we want that top card to be a mile beyond the edge of the table, that would require enough cards to have this sum of fractions be bigger than 40,000. How many cards are needed? The number has approximately 17,000 digits! To put that number in perspective, the total number of atoms in the universe is an 80-digit number. So cards could theoretically be stacked to cantilever out a mile, but you shouldn't try it at home. Certainly this 17,000-digit number dramatically dwarfs all other numbers featured in this chapter—this number truly takes the cake.

Finally, we note that sitting on the cantilevered stack of cards (as in *Figure 5.16*) is now easy. All we have to do to determine where we should sit is to remove a collection of cards from the top—a collection that weighs slightly more than we ourselves weigh—and then sit down on the top remaining card. The center of gravity of those cantilevered cards is at the very end of the next card, so we will be safe sitting serenely on the top of a very high stack of cards—a highly counterintuitive sight to behold.

SUMMING UP

Numbers are truly fascinating objects and come in all shapes and sizes. Thousands, millions, billions, trillions, quadrillions, and beyond can all have meaning to us. To appreciate these magnitudes we can visualize scenarios from counting stars to stacking cards. When we get into the habit of counting, we open our minds to new worlds of precision and nuance.

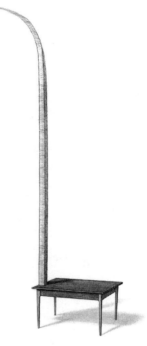

Fig. 5.16

CHAPTER 6

A SYNERGY BETWEEN NATURE AND NUMBER

A Search for Pattern

If we do not expect the unexpected, we will never find it. —Heraclitus

Obviously . . . An 8 × 8 square has an area of 64 square units. Clearly, you cannot cut it into puzzle pieces and reassemble them to create a 5 × 13 rectangle, which has an area of 65 square units (*Figure 6.1*).

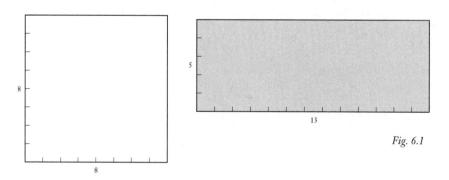

Fig. 6.1

Surprise . . . Consider cutting the square up into the four pieces as illustrated (*Figure 6.2*). Now attempt the feat of making the rectangle out of them. Of course, it can't *really* be done, because the same pieces reassembled would have to have the same area. But you may

Fig. 6.2

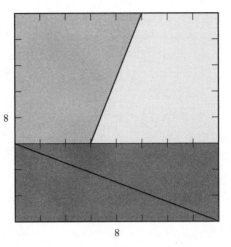

well be able to come up with a very convincing-looking rectangle, one with the merest slit along the diagonal. Why this geometric illusion works is a chapter in the story that starts with a pineapple and ends with fine art.

———

Quite often, not just in mathematics but also in life, if we seek patterns, we find them. And once we uncover patterns, previously invisible features magically emerge, and, suddenly, rich structure appears in laser focus.

PINING FOR PINEAPPLES

Where do we find the inspiration for great discoveries? We might guess that profound insights occur only in well-endowed laboratories or in supple leather chairs surrounded by dusty tomes within the confines of ivy-covered walls. Our story begins, however, in a hum-bler—and dare we say more *fruitful*—setting, the produce aisles of a grocery store. Specifically, we turn our gaze to a tasty tropical fruit—the pineapple.

Pineapples are at once exotic, prickly, and delicious. While normally we focus on the fla-vor they have on the inside, here we consider the physical attractiveness they have on the outside.

We begin by simply *looking* at a pineapple. In fact, we want everyone to acquire his or her own pineapple and explore along with us. As an added incentive—although any hour spent pondering mathematical ideas is a happy hour, consider the colada opportunities that await us after the mathematical features of our pineapples have been distilled. Cheers!

Fig. 6.3a

When we look closely at the pineapple's formidable façade, we soon notice the surprising reality that the face of the pineapple is creased with a collection of spirals. We can run our fingers along the grooves created by the neatly arranged bumps and feel the spiral sweeps that encircle the pineapple. (The hidden spirals are high-lighted in Figure 6.3b.) Now when holding a pineapple and looking at its knobby surface, we will no longer see random bumps. The spiral lines, which were always there but unnoticed, now seem obvious.

Once we discover structure, if we look even closer, we often see even more. In this instance, a second look at our pineapple reveals a second sequence of parallel spirals, these going in the opposite direction (*Figure 6.3c*). These two interlacing collections of spirals mesh together to create the beautiful, familiar pineapple façade (*Figure 6.3d*).

Fig. 6.3b

Fig. 6.3c   *Fig. 6.3d*

COUNTING ON COUNTING

As we have seen in previous chapters, counting often opens our eyes to new insights. Having found spirals all over our pineapple, we might make the effort to move from the qualitative ("There are *many* spirals") to the quantitative ("The *exact* number of spirals is . . ."). Let's count the number of spirals we see in each of the two directions.

Counting spirals is actually more challenging than you'd think. In Figure 6.3e we've made it easier, but we urge you to compare the

Fig. 6.3e

counts here with those of your own pineapple. Or, if you don't care
to make the $3.99 investment, just move from pineapple to pineap-
ple in the grocery store counting away while attracting the attention
of curious bystanders.

Surprise. Most healthy, well-rounded pineapples have spiral
counts of 8 and 13. The first surprise is that the spirals going in one
direction are more numerous than the spirals in the other direction.
The second surprise is that essentially every pineapple possesses
these same spiral counts.

A WHIRLWIND OF SPIRALS

When we open our eyes to spirals, we find them everywhere. Spirals
abound in nature. If we leave the grocery store and stroll into the
garden, a quick glance at the fiery center of a coneflower immediately reveals two sets of interlocking spirals (*Figure 6.4*). How can we

Fig. 6.4

curb our enthusiasm to count them? We can't, so we count (*Figure
6.5a*). In one direction we see 13—an amazing coincidence, given
that one of the spiral counts on the pineapple was 13. We can't stop
now, so we count the spirals in the other direction (*Figure 6.5b*).
Here we move up to 21 spirals.

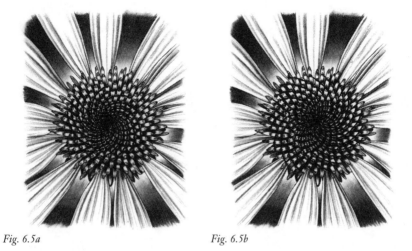

Fig. 6.5a   *Fig. 6.5b*

Thus far the numbers spiraling around in our heads are 8, 13, 21. If we turn to the beautiful daisy, we may have the romantic urge to pluck off the white petals while muttering, "Loves me, loves me not . . ." Whether we succumb to the urge to pluck or not, we are faced with the sunny smile of the circular yellow center consisting of interlocking spirals (*Figure 6.6*). Once we look at familiar objects

Fig. 6.6

very, very closely, a world of structure suddenly unfolds. As we count the daisy's spirals, we see 21 in one direction (*Figure 6.7a*)—again an

Fig. 6.7a

amazing coincidence, since that number agrees with one of our previous counts—and 34 in the other direction (*Figure 6.7b*). Nature's number sequence has grown to 8, 13, 21, 34.

Fig. 6.7b

If we were to pick up a pine cone and count its spirals, we would discover spiral counts of 5 and 8—the 8 providing still another cosmic counting coincidence with the pineapple. This ever-expanding collection of "coincidences" leads us to wonder whether there might be some invisible structure involved. Nature's list of numbers thus far is 5, 8, 13, 21, 34. If we consider this sequence of numbers as a collective whole, an astonishing pattern comes into focus. Do you see it?

Surprise. If we simply add the first two numbers on our list, we produce the next number; the second and third numbers produce the fourth; the third and fourth added together produce the fifth. Amazing—these numbers represent different spiral counts that arise from different fruits and flowers, and yet taken as a group, the numbers on the list exhibit a pattern.

Moving from down-to-earth nature to abstract mathematics, we can certainly continue the sequence of numbers by following the pattern. The next few terms would be 21 + 34, also known as 55, followed by 34 + 55 = 89, then 144, then 233; and we could continue forever. The pattern also invites us to journey backward—that is, before the 5 we must have a 3 (since 3 + 5 = 8), and before the 3 we must see a 2, which must be preceded by a 1 and before that by another 1. Thus we have produced a list of numbers that begins

$$1, 1, 2, 3, 5, 8, 13, 21, 34, 55, 89, 144, 233$$

and continues on forever. This list results from taking a seed of a pattern that we found in nature and then letting it grow in the abstract world of mathematics.

These abstract numbers have implications in the real world. We can shine our sequence of numbers back into the warmth of our everyday world to illuminate new insights into nature. Where would we look? It's natural to consider objects similar to the pine cone, pineapple, coneflower, and daisy from which the sequence of numbers arose. And when we consider the enormous sunflower, we are not disappointed. How many spirals would we see? Given our previous counting coincidences, we might guess 34 and 55, or perhaps 55 and 89. It turns out that these guesses are correct—small sunflowers have counts of 34 and 55, and larger sunflowers have 55 and 89. Count for yourself. Suddenly, invisible structure in our everyday world—connections between seemingly different objects—comes into focus.

Just by looking at objects and issues closely and by taking the refining step of quantifying our observations, we often uncover hidden detail. Usually we see our world in a fuzzy manner, but once we open our eyes and minds to detail and look for patterns, we see a world of richer design and clearer focus.

GROWING ARTIFICIAL FLOWERS

Why do so many of nature's spirals conform to this sequence of numbers? Actually neither mathematicians nor biologists have a complete explanation. It is known that the tiny florets that make up the spirals we were counting—for example, the yellow buds in the daisy—grow from the center. As newer buds emerge, the older buds move out toward the circular edge of the flower. Through this process of growth, every bud is being pushed out to the boundary as the younger little florets appear on the scene.

One theory asserts that these florets, these little, teeny buds, are just like people—they want as much space around them as possible. They want a little plot of land to homestead. Of course, as much as they may try to get some elbow room, the multitude of buds will be squashed in like sardines in a circular can by the time the flower has reached its full size. Scientists have simulated this behavior through computer—rather than sardine—models. That is, they assumed that these little florets are going to grow from the center of a circle and position themselves in such a fashion that they have as much room around them as possible. Once the packing process is complete, the computer models, just like natural flowers, produce images that consist of pairs of collections of spirals whose counts are two adjacent numbers in the sequence of numbers we found. The larger the circular constraint, the larger the spiral counts. These computer simulations do not explain why the spirals arise as they do; however, they do suggest that maximizing the space of each floret lies at the heart of this phenomenon.

REPRODUCTIVE HABITS OF RABBITS

Letting nature ebb and mathematics flow, we note that this list of numbers is referred to as the *Fibonacci numbers*, named after the thirteenth-century mathematician Leonardo de Pisa. Leonardo was also known as Fibonacci, since he was a member of the Bonacci family. In fact, Fibonacci gave himself the nickname Bigallo, which can mean either a much-traveled man or a good-for-nothing. It is well documented that Fibonacci was extremely well traveled and was fascinated with mathematical questions that appeared to have no practical value—so we leave you to ponder the question of which meaning of his nickname Fibonacci meant.

In 1202, Fibonacci wrote a treatise on arithmetic and algebra, entitled *Liber abaci*. In it he posed this question: "A certain man put a pair of rabbits in a place surrounded on all sides by a wall. How many pairs of rabbits can be produced from that pair in a year if it is supposed that every month each pair begets a new pair which from the second month on becomes productive?"

Let's consider his rabbit-generation question. We note that it is assumed, first, that each pair consists of one female and one male rabbit; second, that they are able to procreate after their first month with a one-month gestation period; third, that they have one pair of bunnies each month with precise regularity thereafter; and, finally, that they are monogamous—remember the old expression "There are no bigamists in rabbit holes."

We begin with a pair of baby rabbits at the beginning of the first month. (When referring to a month, we will always mean the beginning of the month.) A month later, that pair of bunnies has become adult rabbits, and the following month they produce their first pair of offspring bunnies.

So the first month we have one pair of bunnies, and the second month we still have one pair, but now they're ready to roll in the hay—"If the cage is a-rockin' don't come a-knockin'." At the third month we see the original, now-proud-parent pair together with their new baby pair.

In the fourth month the original pair gives birth yet again, while the first pair of offspring grows into adulthood—thus we have three pairs. In the fifth month we again have the original couple producing another new baby pair, and now we have the original offspring pair mature enough to reproduce (and they do). However, the second offspring pair merely grows up. So we have a total of five pairs. If we—or, more importantly, the rabbits—carefully continue the process without any timely or untimely demises, then we could produce the family tree shown in Figure 6.8.

Fig. 6.8

Surprise. It's the *Fibonacci sequence* yet again. This rabbit reproductive riddle is the first recorded instance in which these numbers appear, and that is why they are named after Fibonacci. But, of course, nature had us all beat, since she encoded this wonderful sequence in her spirals long before Fibonacci came along in the thirteenth century with his question about bouncing bunnies. It's refreshing to think that if we wait long enough, humankind can eventually catch up with nature, or at least a bit of it. But perhaps we should give nature her due and call the Fibonacci numbers "nature's numbers."

As a postscript, we leave it to you to answer Fibonacci's original question and determine how many rabbits will be hopping around after a year. All we will say is, the enclosed area had better be quite large. Also, if given the choice, you should volunteer to do the feeding rather than the cleanup afterward.

REPRODUCTIVE NUMBERS

Moving from the reproductive habits of rabbits, we turn to the reproductive habits of the Fibonacci numbers. Certainly the farther we travel along the list, the larger the Fibonacci numbers become. But how quickly are they growing? Given that nature offered us this list of numbers in pairs, abstracted from the spiral counts, it seems natural to make a ratio out of each pair of consecutive numbers on our list. Those ratios and their decimal equivalents provide us with a sense of how fast the Fibonacci numbers are growing (*Figure 6.9*).

Fig. 6.9

Ratio of adjacent Fibonacci numbers		Decimal equivalent
$\frac{1}{1}$	=	1.0
$\frac{2}{1}$	=	2.0
$\frac{3}{2}$	=	1.5
$\frac{5}{3}$	=	1.666...
$\frac{8}{5}$	=	1.6
$\frac{13}{8}$	=	1.625
$\frac{21}{13}$	=	1.6153...
$\frac{34}{21}$	=	1.6190...
$\frac{55}{34}$	=	1.6176...
$\frac{89}{55}$	=	1.6181...
$\frac{144}{89}$	=	1.6179...

What is striking as we move down the list of ratios of consecutive Fibonacci numbers is that the decimal equivalents seem to be approaching a particular value. Those numbers are getting closer and closer to one another and heading toward 1.61803*something*, a number that does not, at the moment, appear to be particularly attractive or natural. But perhaps we are not looking at it in the right light.

THE AREA PARADOX PUZZLE. The chapter-opening puzzle paradox actually gives us a visual means of confirming that those successive quotients are indeed approaching some fixed value. If you solved the puzzle, you saw that an 8 × 8 square having an area of 64 could be cut up into pieces and rearranged to form what appeared to be a 5 × 13 rectangle having an area of 65. Thus, by moving the puzzle pieces around, you made the paradoxical discovery that you could increase the area! If this phenomenon were really possible, then your next move would be clear: You'd purchase a tiny plot of land in Hawaii, 1 foot by 1 foot, and then you'd move pieces of it around until eventually you owned the entire island—not a bad investment. Of course, common sense and mathematics tell us that area can't expand this way. You couldn't *really* do what we suggested. What went wrong with the square-to-rectangle transformation?

If we look closely at the image of the rectangle that you probably came up with (*Figure 6.10*), we see that there's a nearly invisible gap running along the diagonal. That slight gap is the extra unit of area that we seem to have gained. In terms of our number list, we notice that 5, 8, and 13—the numbers used for the dimensions of the square and the rectangle—are three consecutive Fibonacci numbers. Note again that 8 × 8 and 5 × 13 differ by 1. It turns out that there is a pattern here: For any three consecutive Fibonacci numbers, the middle number multiplied by itself will always differ from the product of the two flanking numbers by exactly 1. As a further illustration, consider the Fibonacci triple 89, 144, 233. Multiply the middle number, 144, by itself, and compare that amount to the product of 89 and 233:

Fig. 6.10   The shaded sliver has an area of one square unit.

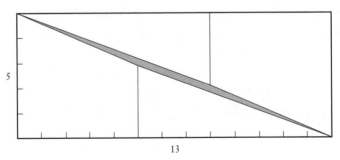

5

13

$$(144 \times 144) - (89 \times 233) = 20{,}736 - 20{,}737 = -1.$$

If we perform some division, we can express this observation in a fancier mathematical manner. In terms of the square and rectangle dimensions of 5, 8, and 13, we see that

$$\frac{8}{5} - \frac{13}{8} = -\frac{1}{5 \times 8} = -\frac{1}{40}.$$

More generally, given three consecutive Fibonacci numbers, the following beautiful relationship is always true:

$$\frac{\text{middle Fibonacci \#}}{\text{first Fibonacci \#}} - \frac{\text{last Fibonacci \#}}{\text{middle Fibonacci \#}} = \pm \frac{1}{(\text{first Fibonacci \#})\ (\text{middle Fibonacci \#})}.$$

This observation implies that as the Fibonacci numbers increase in size, the distance between two adjacent fractions from our chart (*Figure 6.9*) decreases toward zero. But what value are those ratios of nature's spirals approaching? The target, a decimal number of the form 1.61803*something*, still appears utterly arbitrary. However, we will find that this number is as beautiful as the spirals that led us to it. To close in on this claim, we must inch up to the number from the right direction.

ONES OVER ONES. We can write out the fractions from our chart in a manner that relates each fraction to the fraction above it. Thus, starting with any ratio on our list, we can repeatedly divide and conquer our way to the top of the chart (*Figure 6.11*). That is, we can

Fig. 6.11

$$\frac{2}{1} = 1 + 1 = 1 + \frac{1}{1}$$

$$\frac{3}{2} = 1 + \frac{1}{2} = 1 + \cfrac{1}{1 + \cfrac{1}{1}}$$

$$\frac{5}{3} = 1 + \frac{2}{3} = 1 + 1\Big/\left(\frac{3}{2}\right) = 1 + \cfrac{1}{\frac{3}{2}} = 1 + \cfrac{1}{1 + \cfrac{1}{1 + \cfrac{1}{1}}}$$

$$\frac{8}{5} = 1 + \frac{3}{5} = 1 + 1\Big/\left(\frac{5}{3}\right) = 1 + \cfrac{1}{\frac{5}{3}} = 1 + \cfrac{1}{1 + \cfrac{1}{1 + \cfrac{1}{1 + \cfrac{1}{1}}}}$$

express any of the ratios as a "continued fraction"—a fraction within a fraction within a fraction and so forth, of the form "1 plus 1 over 1 plus 1 over 1 plus 1 over. . . ," until we arrive at that last 1 (*Figure 6.12*). In fact, each successive fraction in our chart can be obtained from the previous one by replacing the final 1 with $1 + \frac{1}{1}$.

So consecutive Fibonacci numbers reveal an elegant, algebraic

Fig. 6.12

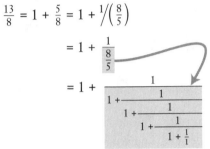

simplicity: We see that the ratio of any two consecutive terms can be written as a continued fraction of long divisions involving only 1's. The natural beauty of spiral counts first drew our attention to consecutive Fibonacci numbers. Now we have found that these consecutive Fibonacci numbers produce a beautiful mathematical structure that rivals nature's spiral counterpart. In fact, this algebraic simplicity is a defining trait. That is, there are no fractions other than ratios of consecutive Fibonacci numbers that can be expressed as continued fractions involving only 1's.

Armed with this new insight, we can put all those 1's into their diagonal formation to home in on the target where all the ratios aim: the number 1.618. . . . Looking at that ever-growing list of 1's, we come to realize that our target must equal 1 plus 1 over 1 plus 1 over 1 plus 1 over 1 plus 1 over 1, *forever* (*Figure 6.13*). That is, the

Fig. 6.13

$$1 + \cfrac{1}{1 + \cfrac{1}{1 + \cfrac{1}{1 + \cfrac{1}{1 + \cfrac{1}{1 + \ldots \text{ forever!}}}}}}$$

unending decimal expression 1.618 . . . can be rewritten as a very simple repeating collection of 1's—with the only wrinkle being that the successive divisions go on forever.

Viewing the number in this new, never-ending manner, we discover an elegant self-similarity. If we simply put a frame around the endless run of 1's (*Figure 6.14*), then what is contained in the frame is another copy of the original number!

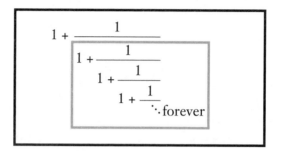

Fig. 6.14   The number inside the gray frame is identical to the entire number inside the large, black frame.

If we call the original number *phi*, the Greek letter φ, then we observe that

$$\varphi = 1 + \frac{1}{\varphi}.$$

Solving this equation for φ gives us a rare opportunity to be reminded of algebra from our high school math daze—all we need is that dreaded "quadratic formula." But instead of taking you through all the uninspiring algebraic steps, we'll simply report that the positive answer is

$$\varphi = \frac{1 + \sqrt{5}}{2}.$$

That square-rooted number may appear threatening and foreign, but if we enter $(1 + \sqrt{5})/2$ into a calculator, the display will show 1.618033989. . . . Our calculator thus confirms that this number is indeed the decimal value that the fractions of adjacent Fibonacci numbers were heading toward.

It's interesting to realize that there is no such thing as a mathematical free lunch. Although we are able to express that endless collection of 1's in a very tight expression, it comes at a price: We must introduce a square root. In some sense, the square root absorbs the infinite complexity of that endless tail of 1's.

The number $(1 + \sqrt{5})/2$ is called the *Golden Ratio*, and, as we will discover in the next chapter, this number has tickled the imagination of artists for millennia. In fact, this number may inform our notion of aesthetics, and we may be led to wonder if perhaps mathematics and our personal tastes are closely entwined.

SOWING SEEDS AND REAPING RECURSION

The endless continued fraction of 1's that yields the Golden Ratio arises from the endless list of Fibonacci numbers. But the Fibonacci numbers themselves can be described in just two steps. First, we start with two initial values, 1 and 1, which we view as starting seeds. And second, we describe the process for generating the next term— namely, we add the previous two numbers. The seeds together with the rule generate the entire Fibonacci sequence. In fact, this sequence is an example of what is called a *recurrence sequence*, because we use the previous values in our sequence to generate the next number. All we need to know are the first couple of values and the process.

When we highlight the essential ingredients of our construction process in this way, we see that we can generalize the Fibonacci numbers and consider even more exotic collections of numbers by simply choosing different starting seeds. For example, suppose we begin with the starting seeds 2 and 1, rather than 1 and 1. If we keep the same generating process but start with 2 and 1, what do we see? The next number would be $2 + 1 = 3$, and then $1 + 3 = 4$, and then $3 + 4 = 7$, and so forth. If we continue, then the new sequence of numbers would look like this:

$$2, 1, 3, 4, 7, 11, 18, 29, 47, 76, \ldots$$

This new sequence also has a name; it's called the *Lucas sequence*, in honor of Eduard Lucas, a nineteenth-century French mathematician who studied recurrence sequences.

The Lucas sequence has very little in common with the Fibonacci sequence. Once we pass the first few terms, we have two lists of numbers that seem completely different.

Fibonacci numbers: 1, 1, 2, 3, 5, 8, 13, 21, 34, 55, 89, 144, . . .

Lucas numbers: 2, 1, 3, 4, 7, 11, 18, 29, 47, 76, 123, 199, . . .

Even though they have the same generating process—add the previous two to get the next one—the lists that are generated appear unrelated, since the starting seeds are different. But are the lists in fact unrelated?

Let's see how the Lucas numbers grow. Of course, we now know how we should measure that growth—by considering the ratios of consecutive terms and their decimal equivalents (*Figure 6.15*). So where do we land?

Fig. 6.15

$$\frac{1}{2} = 0.5$$

$$\frac{3}{1} = 3.0$$

$$\frac{4}{3} = 1.333\ldots$$

$$\frac{7}{4} = 1.75$$

$$\frac{11}{7} = 1.5714\ldots$$

$$\frac{18}{11} = 1.6363\ldots$$

$$\frac{29}{18} = 1.61111\ldots$$

$$\frac{47}{29} = 1.62068\ldots$$

$$\frac{76}{47} = 1.61702\ldots$$

$$\frac{123}{76} = 1.6184\ldots$$

Surprise. We seem to be heading toward the Golden Ratio! Can we confirm this conjecture for ourselves? Of course—and we now know exactly how to proceed. We write the ratios of consecutive Lucas numbers as continued fractions and see where they lead us (*Figure 6.16*). We soon discover a continued fraction pattern very

Fig. 6.16

$$\frac{3}{1} = 3$$

$$\frac{4}{3} = 1 + \frac{1}{3}$$

$$\frac{7}{4} = 1 + \frac{3}{4} = 1 + 1/\left(\frac{4}{3}\right) = 1 + \frac{1}{\frac{4}{3}} = 1 + \frac{1}{1 + \frac{1}{3}}$$

$$\frac{11}{7} = 1 + \frac{4}{7} = 1 + 1/\left(\frac{7}{4}\right) = 1 + \cfrac{1}{1 + \cfrac{1}{1 + \cfrac{1}{3}}}$$

$$\frac{18}{11} = 1 + \frac{7}{11} = 1 + 1/\left(\frac{11}{7}\right) = 1 + \cfrac{1}{1 + \cfrac{1}{1 + \cfrac{1}{1 + \cfrac{1}{3}}}}$$

$$\frac{29}{18} = 1 + \frac{11}{18} = 1 + 1/\left(\frac{18}{11}\right) = 1 + \cfrac{1}{1 + \cfrac{1}{1 + \cfrac{1}{1 + \cfrac{1}{1 + \cfrac{1}{3}}}}}$$

similar to the one we generated with Fibonacci numbers. But here, instead of seeing only 1's, we now see 1's except for the very last number, which is always a 3. But what would happen if we repeated the process forever, as required to find the limiting target? That last 3 would keep drifting farther and farther and farther off toward infinity, and if we repeated the process *forever*, the 3 would disappear off the horizon. We'd be left with an endless stream of 1 plus 1 over 1 plus 1 over 1 plus 1 over 1 plus 1 over 1, *forever*—which brings us to the Golden Ratio.

So both roads lead to the Golden Ratio! This Lucas sequence discovery actually illustrates a profound strategy for understanding—seek the essential. Which features of a situation are truly important and which are irrelevant? We have seen that the ratios of consecutive Fibonacci numbers lead to the Golden Ratio and that the ratios of consecutive Lucas numbers also lead to the Golden Ratio. By looking at the similarity of the arguments, we can see a far more general truth as well.

Suppose we begin with *any* two nonzero starting numbers and produce an endless list of numbers by the rule of adding the two previous terms in order to generate the next number on the list. Of course, different starting seeds will yield entirely different sequences. However, no matter what the two starting seeds are—pick them big, pick them small—the ratios of consecutive terms of the sequence will approach the Golden Ratio. Thus, all such recurrent roads lead to the Golden Ratio. The starting values are irrelevant; the process is everything. We have isolated the essential and identified the irrelevant.

SUMMING UP

Nature led us to a natural list of numbers, the Fibonacci numbers, and they, in turn, led us to the Golden Ratio. That number first appeared arbitrary and foreign. However, after following the patterns and looking at simple things deeply, we discovered that the Golden Ratio has surprising hidden beauty and structure and that it is far from arbitrary. For within every number sequence we form by adding the two previous terms to generate the next, the ratios of consecutive terms fly to the Golden Ratio as moths to the flame.

Nature and mathematics surprise us with unexpected synergy. We found the Fibonacci numbers in different flowers and fruits, and we found the Golden Ratio in unexpected sequences. In the next chapter, we will turn our minds and our sights to the visual world of geometry, and there we will find reflections of the Golden Ratio both in the natural world and in the world of art.

PART III

EXPLORING AESTHETICS

Sexy Rectangles, Fiery Fractals, and
Contortions of Space

When we open our eyes at the break of dawn, we are greeted by the varied display of our visual world. Whether it dawns on us or not, opening our eyes to the geometrical possibilities of our world adds richness and wonder to our lives. Our sojourn into geometrical ideas offers us three glimpses into visible splendor: the classical beauty of the Golden Rectangle; the infinitely complex gracefulness of fiery fractals; and the surprising possibilities that arise when we imagine our world as contortable and elastic. The physical and geometrical aspects of experience let us indulge our urge to see, feel, and delight in the shapes and forms that bring us beauty.

We begin with one of the simplest forms we can construct, the modest rectangle. Among its rectangular brethren, the sexiest among these pillars of quadrilateral rectitude is the Golden Rectangle. Its place in the history of aesthetic ideas is indisputable. Artists such as Mondrian and architects such as Le Corbusier have deliberately injected this most attractively proportioned rectangle in their works. This seductive rectangle is not without controversy—we will find that it has such allure that sometimes people see it in places where it may not even exist. We will uncover what makes the Golden Rectangle so golden and wonder whether a self-replicating feature might explain its omnipresence throughout the ages.

The masters of the art of origami create graceful swans, dragons, and airplanes. Rather than try to imitate the masters with their delicate skills, we will always make the simplest possible folds in our exploration of this art. However, we will allow ourselves the mathematical luxury of abandoning reality and imagining numbers of folds well beyond those that are physically possible. What we will find are patterns and possibilities that change the apparent chaos into beautiful order. In the simplest sequence of paper folding, we find the complexity of the most complicated computers and the fiery complexion of the fractal Dragon Curve. Complexity and structure often grow out of a simple process repeated with infinite persistence.

We complete our tour of the visual world by imagining and then exploring a world made of unreasonably elastic rubber. This imaginary world, though related to our own, allows bendability far beyond reality. Thinking about distorting one object into another opens our minds to truly surprising possibilities. Amazingly, for example, a self-linked two-holed doughnut can be stretched, without any cutting or tearing, into an unlinked version. We will discover that through the tomfoolery of playing with imaginary rubber, we gain insight into our actual, more rigid world, including new understanding of the reproductive behavior of DNA.

These glimpses into the Golden Rectangle, paper folding, and rubber-sheet geometry all show us a way of parsing a complex experience and finding pattern and order that explain or describe it. Looking for patterns and letting our minds explore worlds that are not entirely possible often serve to illuminate our real world with a richer light.

———————

CHAPTER 7

FROM PRECISE BEAUTY TO PURE CHAOS

Picturing Aesthetics Through the Lens of Mathematics

The mathematical sciences particularly exhibit order, symmetry, and limitations; and these are the greatest forms of beauty.—Aristotle

Obviously . . . The Parthenon in Greece, Debussy's musical work *Prelude to the Afternoon of a Faun*, and the position of your bellybutton have absolutely nothing in common.

Surprise . . . The alluring Golden Ratio unites the Parthenon and Debussy, and it possibly pinpoints the perfect placement for our bellybuttons. If we embrace the Platonic ideal of beauty and form, then we are faced with the real possibility that mathematics informs our aesthetic tastes and determines what we perceive as attractive.

―――――

PROPORTIONS WITH PANACHE—
THE SEXIEST RECTANGLE

Turn on some romantic music, lie back on a loveseat, close your eyes, and imagine the ideal rectangle—neither too square nor too elongated. With that rectangle dancing in your head and tickling your fancy, consider the four rectangles shown in Figure 7.1 and determine which of these is closest to the one that floated into your consciousness.

Fig. 7.1

Surprise. Did you pick the third rectangle from the left? Many people fantasize about that one. We do confess that if we actually held an election, it would probably not be a landslide. In fact, in electing the winner of the rectangular prize, we might well relive the 2000 nightmare of hanging-chad issues in Florida. However, a plurality tends to be drawn to that third rectangle—avoiding the need for the Supreme Court to rule on *Base* v. *Height.*

Rectangles, just like people, come in many shapes—tall and thin, short and wide, and everything in between. But some rectangles are so elegantly proportioned that we call them Golden Rectangles. Their proportions are breathtaking to behold. Such a rectangle is the quintessence of rectangularity, the *sine qua non* of rectangleosity, the sexiest rectangle ever. The proportions that generate this hyperbole of effusion are intimately connected with the Golden Ratio— the number that arose organically from pine cones, pineapples, and daisies.

In Chapter 6 we discovered that nature's spirals are not unlike the cabin dwellers on Noah's Ark: They come in pairs—one number for the spirals in one direction, the other number for the spirals in the other direction. Those spiral pair counts fit together to form consecutive terms in the list of Fibonacci numbers: 1, 1, 2, 3, 5, 8,

13, 21, 34, 55, and so forth. The ratios of consecutive terms in this idealized spiral-count sequence approach the supremely attractive Golden Ratio, denoted by the Greek letter phi (φ) and numerically equal to 1.618. . . .

A *Golden Rectangle* is a rectangle for which the ratio between its base and its height is the Golden Ratio. The third rectangle depicted in Figure 7.1 is an example of a Golden Rectangle. Rectangles with these proportions entice us at every turn. Haven't you noticed an inexplicable desire to use index cards? They're everywhere. Find one and measure its dimensions: 3×5 if it's a standard issue, or 5×8 if you are in possession of a nice big one.

The ratios of base to height of those seductive index cards are $5/3 = 1.666.$ . . and $8/5 = 1.600.$ . . , two values that approach the now-familiar digits 1.618. . . that compose the digital DNA of the Golden Ratio. Perhaps this fact is not too surprising, since both (3, 5) and (5, 8) are pairs of consecutive Fibonacci numbers. Maybe those friendly folks at the ad agencies on Madison Avenue carefully picked those dimensions as a subliminal message so that we will feel drawn to buy index cards. We do go through a good number of them.

While a connection between mathematics and Madison Avenue through the proportions of an index card seems as flimsy as the index card itself, there are many real instances where we see reflections of the Golden Rectangle in artistic works throughout human history. We begin with the ancient Greeks.

GOLDEN OLDIES

The Parthenon is a supreme example of classical beauty, but we probably would not be thrilled with our on-line virtual travel agent if that's where it arranged for us to stay during a visit to Athens. Let's face it, the place is in ruins (*Figure* 7.2). But it was not always so. Once, the Parthenon was the "in" place to be—"in" in the sense that it had a roof, anyway. If we extrapolate from the remaining roofline fragments, we can envision the Parthenon as it looked in its heyday, and then we can superimpose a Golden Rectangle that almost per-

Fig. 7.2

fectly encloses the newly shingled ancient edifice (*Figure* 7.3). Is the Parthenon's connection with the Golden Rectangle by design or purely a coincidence?

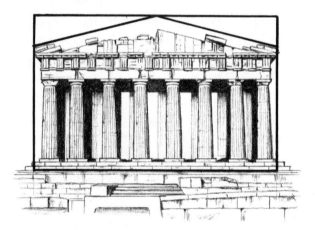

Fig. 7.3

Before considering the design-versus-coincidence question, we eye another example from ancient Greek culture. Long before disposable contact lenses, the ancient Greeks required objects known as Grecian eyecups to apply Visine—well, probably plain water—to their ancient Grecian eyes. These decorated cups were often adorned with a pair of eyes peering back at the beholder, leaving no doubt about their intended purpose. As we admire the sweet grace of these morsels of eyecup eye candy, we find Golden proportions staring us in the face (*Figure* 7.4). If we declare the distance from the far end of one handle to the edge of the cup to be equal to $\frac{1}{2}$, then the height of the cup would be surprisingly close to the Golden Ratio. Moreover, the length of the opening of the eyecup would equal $1 + \varphi$.

Fig. 7.4

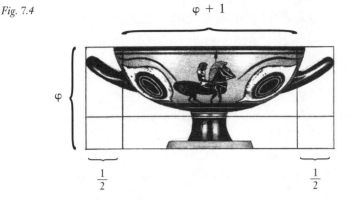

Was the use of the Golden Ratio and the Golden Rectangle in these ancient works of art intentional? This question remains open to this day. A romantic conjecture is that the Greeks subconsciously selected those proportions on the basis of an innate sense of grace and beauty, thus wedding intrinsic aesthetic appeal with abstract mathematical constructs.

DIVINE PROPORTIONS

Fig. 7.5

Fast-forwarding a millennium and a half or so brings us to the Renaissance. Leonardo da Vinci was well aware of the Golden Rectangle and the Golden Ratio; in fact, he was the illustrator for a treatise on the subject entitled *De Divina Proportione*. We can find the divine Golden Ratio appearing gracefully in his art. If we study his unfinished portrait of Saint Jerome, for example, we see that we can draw a perfect Golden Rectangle that surrounds the saint's body (*Figure 7.5*). However, we do not know whether Leonardo intentionally crafted the proportions of this seated saint to conform to the Golden Rectangle or whether sheer aesthetic taste dictated the balance of the figure.

Actually, a more fundamental and controversial question is whether the Golden Rectangle proportions are even there at all. If we look closely at that superimposed Golden Rectangle, we notice something a little fishy—the left edge of that rectangle doesn't touch the figure or even the lower portion of the draped cape. That edge is just hanging out there for no apparent reason except to make the rectangle perfectly Golden.

In fact, we can lose even more Golden ground if we return to the Parthenon and examine the lower edge of the superimposed rectangle. Now we notice that the Golden Rectangle is resting randomly on the second step. So is the Golden Rectangle actually there at all? While to those members of the cult of the Golden Ratio such observations are heresy, in reality no perfect Golden Rectangle can ever be built, since we are unable to measure things with infinite precision. Nevertheless, some influences of the Golden Rectangle on art are indisputable.

Continuing our whirlwind tour through the history of art, we arrive at the Impressionist era in France. At that moment in history, artists turned to scientific ideas for inspiration. Georges Seurat was an artist who deeply embraced this interplay between mathematical ideas and art. In fact, some have said that Seurat attacked every canvas with the Golden Rectangle in mind (although we note that others have disagreed). In *La Parade*, we find many Golden Rectangles (*Figure 7.6*). Moreover, he constructed the ratios of various lengths in his works that seem to closely exhibit the Golden Ratio. Are they there or not? What do you think?

If we jump to Expressionism and Modernism, we might well be confronted with blank canvases or, if we're lucky, rectangles. Piet Mondrian was an exponent of Expressionist expression known for his concept of "destruction of volume," which refers to space being cut up into pieces. In a pastime that has much in common with the child's game of searching for the hidden Waldo in a sea of near look-alikes, we can search Mondrian's paintings for Golden Rectangles possibly hidden among the other lines and shapes. It is said that

Fig. 7.6   Three Golden Rectangles (corners marked by ①'s, ②'s, and ③'s).

when a viewer commented on all the lines in his work, Mondrian responded, "I don't see any lines at all. All I see is the space surrounding them." Of course, the space is where those Golden Rectangles may reside.

CONSTRUCTING AND COMPOSING GOLDEN WORKS

Unlike most of the general public, the twentieth-century French architect Le Corbusier believed that human life is soothed by mathematical ideas. Among the most soothing, from his point of view, was the Golden Rectangle, which he intentionally incorporated in many of his works. In Figure 7.7 we see his French villa, which, if we include the chimneys, can be encased in a nearly perfect Golden Rectangle. In fact, as we'll soon discover, Le Corbusier's aesthetically appealing environment actually captures a feature possessed only by our glamorous Golden Rectangle.

Our appreciation for the Golden Ratio need not be confined to the visual world. In fact, we can sit back and *listen* to the beauty of that seductive value. Claude Debussy was fascinated with the Golden Ratio and strove to capture that number in his musical works. In particular, in his *Prelude to the Afternoon of a Faun*, if we lis-

Fig. 7.7

ten carefully, we can hear the Golden Ratio calling to us in pitch, rhythm, and dynamics.

Musical pulses are known as *quaver units*. If we measure the quaver units of *Prelude to the Afternoon of a Faun*, we notice some interesting patterns. Debussy notated *fortissimos* (*ff*, very loud) at bars 19 and 28. If we add these two values, we find ourselves at the next *fortissimo*, at bar 47. Thus we see reflections of the Fibonacci pattern in Debussy's musical work (*Figure 7.8*). In fact, examining the musical dynamics illustrated in the figure reveals many other such Fibonacci patterns.

The piece builds to a dramatic *fortissimo* at bar 70 and then gradually moves to *pianissimo* (*ppp*, very quiet). The length of the entire prelude is 129 seconds, and the length of time between the beginning of the piece and that fantastic *fortissimo* at bar 70 is 81 seconds. Dividing those two quantities yields 1.592. . . , which is impressively close to the Golden Ratio of 1.618. . . . Is part of the beauty and elegance of Debussy's work due to the intentional appearance of the Golden Ratio? Although we can't be sure of that, we *can* be sure that beauty is all around us—and that once we become attuned to it, we can appreciate it with our eyes, ears, and mind.

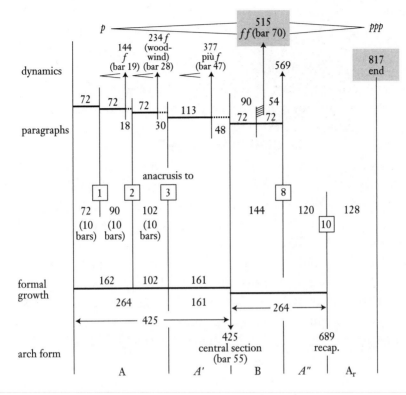

Fig. 7.8   *Quaver units for Debussy's work. Note: 144f + 234f = 377f.*

MINING FOR GOLD

Given our current infomercial culture, after all this Golden Rectangle hype, you can almost hear the pitchman yell out to the paid studio audience, "Well, I'm sold—I think we all want one of our own, don't we?" To which the audience would applaud wildly, as specified in their contracts. Although dialing 1-800-GoldRec will not reach an operator who is eagerly standing by to take your order for the rectangle of your dreams, we can pose the following materialistic question: Is there a process by which we can construct our very own copy of a Golden Rectangle?

Actually, a simple geometric process that dates back to the ancient Greeks yields a perfect Golden Rectangle. Ancient Greek mathematicians (actually, they only become ancient a thousand or so years after their deaths) were fascinated with the question of which geometric objects could be constructed using only a straightedge and a compass. Of course this was long before cable TV, so at the time this question was fascinating. But seriously, straightedge and compass constructions have engaged people for centuries and will continue to do so for much longer than, say, *Leave It to Beaver.* Most of us have firsthand experience with such constructions from back in our high school geometry classes.

Recall from your youth that classical geometric constructions specify that we use a straightedge that is unmarked. That is, the edge is not a ruler. Perhaps the rule against rulers is appropriate, since nothing in our physical world can be measured exactly. Just as we ourselves discovered a few chapters ago with two working calculators producing two dramatically different answers, any perfect measurement theoretically would require infinitely many digits of accuracy —which is, of course, impossible if we stick to reality. So the idealized mathematical construction challenge that we undertake is to use only a straight, unmarked edge and a compass and to perform a geometric construction that produces a perfect Golden Rectangle. Here are the steps:

Step 1. Our construction begins with a perfect square. (Given any line segment, we can use a straightedge and a compass to build a perfect square, one of whose sides is the given line segment. We can use those same two tools to pinpoint the precise middle point of that bottom side. These construction tasks are left as challenges to the geometric purists among our readers.) Next we use the straightedge to extend the bottom edge of the square off to the left so we have what appears to be a square house with a driveway off to the left (*Figure 7.9*).

Fig. 7.9

Step 2. Next we place the dangerous (sharp) end of the compass at the midpoint of the bottom edge of the square and the pencil end of the compass at the upper left corner of the square. Given the current compass position, we now sweep out a circle having its center at that midpoint and passing through that upper left corner. Actually we need only a portion of the circle (*Figure 7.10*).

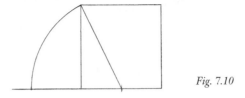

Fig. 7.10

Step 3. By moves similar to those used to construct the starting square, we can create a line that is perpendicular to our driveway and passes through the point of intersection of the driveway and the circle.

Step 4. Finally, we extend the top edge of the square to meet the new perpendicular line, and we can erase extraneous lines and arcs (*Figure 7.11*).

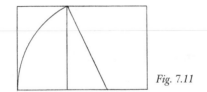

Fig. 7.11

Let's look at what we have constructed. Are you experiencing any rectangular arousal? If so, there is no need to be ashamed—it's perfectly natural, since we have just built a perfect Golden Rectangle. But is it really Golden? Confirmation requires a mathematical proof. As John Lennon once almost wrote, "All we are saying is give proof a chance." And so in his honor we will.

PRECIOUS PROOF

How do we show that a rectangle is Golden? We must show that the ratio of its base to its height is exactly the Golden Ratio, $(1 + \sqrt{5})/2$.

Since we have no scale in our figure, let's simply declare that each of the sides of the square has a length of 2. Thus we instantly see that the height of our rectangle is 2. But what about its bulging base? Unfortunately, that length is not nearly as apparent. To make that length measure up, we must return to the intermediate steps in our construction. If we consider the diagonal line—namely, the radius of the circle that contains the upper left corner of the square—then the underlying structure becomes clear.

First, we observe that the length of that radius is equal to the length of the portion of the base from the midpoint of the square to the left edge of the rectangle. The remaining rightmost length of the base equals 1, since it is half of the base of the 2×2 square. Therefore, instead of finding the length of the leftmost portion of the base of the rectangle, we consider the equivalent issue: What is the length of that new diagonal-esque line? Toward that end, we notice that our construction created a special right triangle with our line as the triangle's hypotenuse. It is easy to see that one leg (the base) is of length 1 and the other (the height) is of length 2 (*Figure 7.12*).

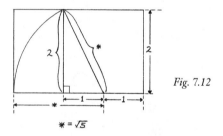

Fig. 7.12

Applying the good old Pythagorean Theorem—the square of the hypotenuse equals the square of the base plus the square of the height—we find that

$$\text{hypotenuse}^2 = 1^2 + 2^2 = 1 + 4 = 5.$$

So the hypotenuse must have a length of $\sqrt{5}$—a now-familiar piece of the Golden Ratio. If we add on the rightmost edge of the base, we see that the exact length of the entire base of the rectangle is $1 + \sqrt{5}$. We now divide this quantity by the height, 2, to discover that the ratio of base to height equals $(1 + \sqrt{5})/2$, and hence we confirm that our rectangle is indeed a Golden Rectangle.

The key to unlocking the idea behind this argument is that critical triangle—a triangle that we will revisit at the end of this chapter for a somewhat chaotic close.

WHY THE APPEAL?

Why do we see proportions conforming to the Golden Ratio in so many works of art? With the experience of our geometric construction under our belts, we return to the villa designed by Le Corbusier. There he overtly included the Golden Rectangle, which we now see in a new, informed light. In particular, our construction method highlighted a natural connection between a Golden Rectangle and the largest square it holds within it. When we look at Le Corbusier's villa, we cannot help but notice that, similarly, the indoor living space on the right creates a large square (*Figure 7.13*).

Fig. 7.13

Thus within Le Corbusier's design we actually find the same geometric construction of the Golden Rectangle that we just outlined: We start with the square interior portion of the villa, and then we adjoin the open patio on the left to create the rectangular appeal of the villa in its totality. Is there anything special about that rectangular patio that was required in order to fill out the Golden Rectangle? If we focus on it, turn it on its side, and enlarge it, we see an unexpected coincidence: That rectangle appears to be another Golden Rectangle (*Figure 7.14*). Actually this architectural observation is a mathematical fact. If we remove the largest square contained in a Golden Rectangle, then the remaining smaller offspring rectangle

Fig. 7.14   *The patio enlarged and placed on its side.*

is another Golden Rectangle. Even more satisfying, this beautiful mathematical truth follows from the beautiful continued fraction representation we discovered for φ in the previous chapter.

THE GOLDEN ONES

Let's convince ourselves that when we remove the largest square possible from a Golden Rectangle, what remains is another Golden Rectangle. We begin with a Golden Rectangle having a height of 1. By definition, the base must be φ, since the ratio of base to height must equal φ. Thus, if we remove the largest square possible (*Figure 7.15*), then we can easily find the dimensions of the smaller offspring

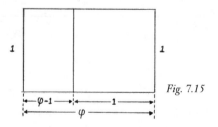

Fig. 7.15

rectangle: 1 by (φ −1). To verify that this smaller rectangle is indeed Golden, we need to prove that the length of the longer side divided by the length of the shorter side is the Golden Ratio, φ. We see that the length of its longer side divided by the length of its shorter side is $1/(\varphi - 1)$, and we wish to show that that value actually equals φ. What does that complicated quotient $1/(\varphi - 1)$ equal?

We return to the beautiful pattern

$$\varphi = 1 + \cfrac{1}{1 + \cfrac{1}{1 + \cdots}}$$

If we subtract 1 from each side of this equality, we see that

$$\varphi - 1 = \cfrac{1}{1 + \cfrac{1}{1 + \cdots}}$$

and so, if we invert both sides to get $1/(\varphi - 1)$, we discover that

$$\frac{1}{\varphi - 1} = 1 + \cfrac{1}{1 + \cfrac{1}{1 + \cdots}}$$

which is precisely φ! Thus we see that the base-to-height ratio of the smaller rectangle is the Golden Ratio, and hence that the offspring rectangle is a baby Golden Rectangle.

In fact, the Golden Rectangle is the *only* rectangle that possesses the property that after the largest possible square is removed, the remaining rectangular offspring has the same proportions of length to width as the original parent rectangle has. Here we have isolated an essential feature that only the Golden Rectangle possesses. Perhaps this unique rectangular similarity illuminates why the Golden Rectangle is so aesthetically pleasing.

AN ENDLESS RECTANGULAR FAMILY TREE

One of the many life lessons that mathematics offers us is that once we discover an idea, we often develop new insights by taking our idea to extremes. In this case we discovered that when the largest possible square is snipped away from a Golden Rectangle, the remaining rectangle is also a Golden Rectangle. So now what? We can repeat! We can now snip off the largest square from our baby Golden Rectangle. What remains? An even smaller grandchild Golden Rectangle. Of course, we could continue this process as long as we wish and thus produce an endless collection of ever-shrinking Golden Rectangles (*Figure 7.16*).

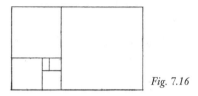

Fig. 7.16

If we take our image and place a quarter-circle within each square, then we create a beautiful spiral (*Figure 7.17*). This spiral is a

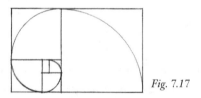

Fig. 7.17

close approximation of a *logarithmic spiral*, the mathematical abstraction of the natural curve exhibited in the graceful nautilus shell (*Figure 7.18*). Yet again we see how mathematical discoveries have informed our understanding of nature and have enhanced our appreciation of artistic beauty.

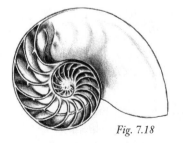

Fig. 7.18

FOOL'S GOLD

While we have seen many illustrations where the Golden Rectangle has appeared in art and nature both intentionally and organically, it is equally true that, no matter how much we fantasize, not all rectangles are golden. Despite this realization, some people have tried to find them even when they're not there.

Let's return to the world of art and study an ink drawing of Le Corbusier's from 1946 called *Modular Man* (*Figure 7.19*). Here we see his attempt at deconstructing the human form into Fibonacci-type patterns. The numbers he lists at the bottom of his piece represent height measurements of various body parts. Starting from the bottom of that number list and reading upward, we see 6 and then 9. If we add those two, we get the next number on his list, and so on. Thus we see a Fibonacci pattern—in fact, from what we discovered in the last chapter we see that this is an example of a recurrence

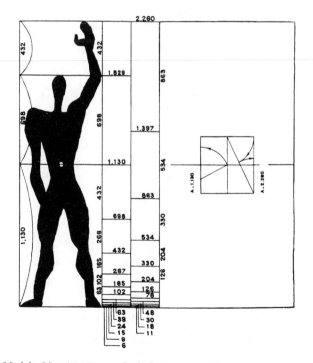

Fig. 7.19   In Modular Man *(1946), we see Le Corbusier creating Fibonacci-esque patterns: 6 + 9 = 15, 9 + 15 = 24, and so forth.*

sequence with starting seeds 6 and 9. As we saw there, the ratio of consecutive terms on this list will approach the Golden Ratio.

The bellybutton on Le Corbusier's drawing reminds us of an amusing myth. Some people believe that in the "ideal" human form, an individual's height divided by the distance of his or her bellybutton from the ground produces the Golden Ratio (*Figure 7.20*). In the privacy of your own home, feel free to measure yourself and see how close you come to being ideal. We do caution that this theory has never been scientifically or mathematically proven, so don't despair if you come up short.

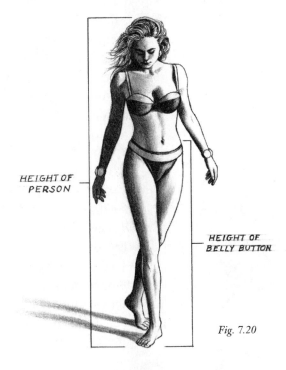

HEIGHT OF PERSON

HEIGHT OF BELLY BUTTON

Fig. 7.20

Some believe that the rectangular dimensions of the palm of one's hand conform to the Golden Rectangle. Again, feel free to trace your hand and measure to see if your palm captures some mythical Platonic aesthetic appeal. Do we see the Golden Rectangle around us or not? Is it in the Parthenon? Is it in the position of our bellybuttons? Perhaps we are literally holding the Golden Rectangle in the palms of our hands. Or perhaps not.

A GOLDEN TRIANGLE

The key player in confirming that our geometric construction did indeed produce a perfect Golden Rectangle was not a rectangle but rather a *triangle*. In fact, it was a right triangle with one leg twice as long as the other. Here we call such a figure a *Golden Triangle*, since it is so central to the geometric construction of the Golden Rectangle (*Figure 7.21*). The Golden Triangle also has unique features of its own that challenge our aesthetic tastes. Thus we close this chapter with a brief exploration of the story of the Golden Triangle and see how it leads to a remarkably hypnotic image of orderly chaos.

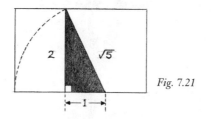

Fig. 7.21

Up to this point, this chapter celebrated the rectangular form and ignored the nuances and three-angled beauty of the triangle. An ordinary, everyday triangle with no special lengths to its sides and no special angles may at first not seem that exciting, but it does have an interesting property: If we make four copies of that triangle, then those triangles can be assembled in a straightforward manner to produce a triangle having the same proportions as the first, but on a larger scale. This larger similar triangle is, in fact, twice the size of the original—that is, each edge is twice as long as the corresponding edge on the original triangle (*Figure 7.22*).

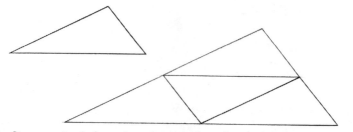

Fig. 7.22   Given any triangle, four copies can be arranged to make a similarly proportioned triangle (twice as long all around).

The Golden Triangle, though, possesses a one-of-a-kind feature that all other triangles do not—a feature that, many people believe, makes it the most aesthetically appealing of triangles. The Golden Triangle is the only right triangle for which *five* copies can be assembled to produce a larger copy of itself (*Figure 7.23*). How? The smaller leg of our large triangle is precisely the hypotenuse of the original right triangle, and the larger leg is made from two hypotenuses. Thus we see that our big right triangle has one leg twice as long as the other, and hence is itself a Golden Triangle. Using this intriguing regeneration process, we can take five copies of any Golden Triangle and produce a larger Golden Triangle—and then of course we can take five copies of that larger Golden Triangle and put the pieces together to form an even larger Golden Triangle.

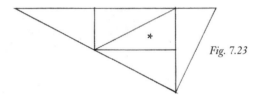

Fig. 7.23

Can we stop now? Of course not. Mathematical patterns are like potato chips—we can't stop at one or two iterations. We all know where iterating the eating of potato chips leads—an expanded waistline. Similarly, we see that iterating this Golden Triangle reproductive process leads to an ever-expanding collection of Golden Triangles, all in need of exercise. But what if we said "Weight Watchers be damned!" and let those Golden Triangles grow *forever*?

REPRODUCTIVE CHAOS

If we repeat the ordinary regeneration process involving only four copies of any triangle, we produce a tiling of the entire plane—that is, an endless floor (*Figure 7.24*). Thus, if we have infinitely many copies of an arbitrarily proportioned triangular tile, we can put them together to build larger but similar versions of the original tile. Repeating the process forever leads to a tiling of the entire plane with tiles of one shape.

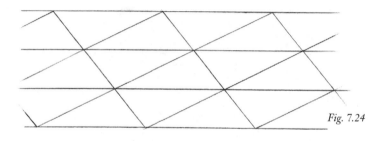

Fig. 7.24

When we study the pattern generated by this tiling, we see a soothing repetition in the pattern that is all at once comforting and attractive. We are actually subconsciously detecting an enormous amount of symmetry present in that tiling. If we shifted the tiling over by one triangle, the new tiling would exactly coincide with the original tiling. We can shift our tiling in many directions and still have the repositioned tiling line up perfectly with the original. We say that this tiling has *translational symmetry*—we can translate, or slide, the entire picture in certain directions, and nothing changes.

If we now take out a massive loan and purchase infinitely many Golden Triangular tiles, then we can tile the entire plane by repeatedly building super-Golden-Triangular tiles. That is, we begin by placing five tiles together to build a super-tile. We can then assemble five copies of this super-tile to construct a super-super-tile that remains a Golden Triangle (*Figure 7.25*). We can repeat this process. When we do so, the tiling that results is surprisingly chaotic or jumbled-looking (*Figure 7.26*).

In fact, it can be shown that there is *no* translation of this tiling that will result in that shifted version matching up perfectly with

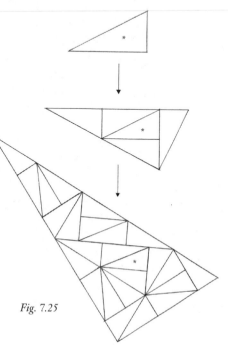

Fig. 7.25

Fig. 7.26

the original tiling. This tiling, known as the *Pinwheel Tiling*, is aperiodic, which means that it has no repeated pattern that would allow us to re-create the pattern after a translation. The inspiration for the name Pinwheel Tiling is the fact that the increasingly larger supertiles that arise during the construction spiral around and point in essentially every possible direction. This chaotic or jumbled image does possess an attractive dissonance. While it is a dramatic departure from the classical beauty of the Golden Rectangle, the Pinwheel Tiling is a modern form of mathematical art that will perhaps inform our aesthetic tastes in the future.

OUR GOLDEN VOYAGE

Our journey into aesthetics and our Golden Rectangular antics bring us to a dramatic realization—not only about spirals, art, and music, but about our personal possibilities for seeing the world more clearly. We have all seen features of nature thousands of times without noticing the detail and rich structure that represents a unifying force. Using simple techniques of thought—in this chapter through seeking the essential regenerative property of the Golden Rectangle—we see the world differently. We see patterns.

How much structure is hiding all around us waiting to be discovered? A story unfolds whenever we look at something closely in the intense light of focused simplicity. By being attuned to detail we suddenly become conscious of a world of beauty illuminated by mathematical pattern. We now see that aesthetics and mathematics are deeply related. There's beauty in mathematics and mathematics in beauty.

CHAPTER 8

ORIGAMI FOR THE ORIGAMICALLY CHALLENGED

From Paper Folding to Computers and Fiery Fractals

> *Not chaoslike, together*
> *crushed and bruised,*
> *But, as the world*
> *harmoniously confused;*
> *Where order in variety we see,*
> *And where, though all things*
> *differ, all agree.*

—Alexander Pope

Obviously . . . The process for creating the infinitely complex image known as the *Dragon Curve* (*Figure 8.1*) requires infinitely complex mathematics.

Surprise . . . The process involves only the mindless task of folding a sheet of paper. The only wrinkle is that we need to repeat that task infinitely many times. When we look for patterns in simple pleas-

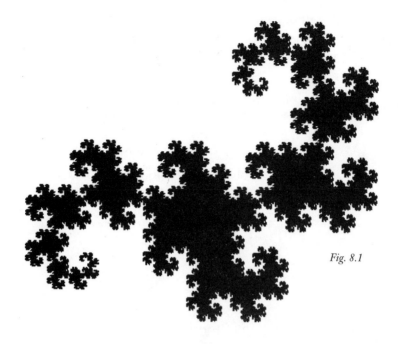

Fig. 8.1

ures, hidden structure leaps out at us—in this case, bringing us face to face with the fiery Dragon Curve. Amazingly, that infinitely corrugated curve can tile the floor with no gaps or overlaps! Furthermore, within the folds of our sheet of paper, we also find the basic foundations of modern computing.

———

Let's be honest—in life we are frequently faced with circumstances that are ridiculously complicated and seemingly impenetrable. Our first impression may be that the situation is hopeless and often ulcer-inducing. In this chapter, we use a journey through the intriguing ups and downs of paper folding as a metaphor for the unraveling of life's complexity. May this journey offer hope to the hopeless and antacid to the ulcerated.

Our main goal here is to embrace the reality that we can understand complex phenomena by looking at simple cases in a deep manner. Through the process of folding a piece of paper repeatedly, we will discover that objects and issues that at first appear to be completely chaotic and devoid of coherence, in actuality possess beautiful structure that leads to wondrous connections among diverse notions. Let's grab a sheet of paper and see what unfolds.

WORKING INTO THE FOLD

We begin with a sheet of paper and the simplest possible move: We fold it in half—right over left (*Figure 8.2*). Even those who are

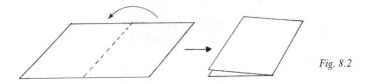

Fig. 8.2

origamically challenged (and incapable of transforming a piece of paper into a graceful swan) can perform this maneuver. However, we will not be content with merely one fold. Instead, we take our folded paper and fold it again—right over left (*Figure 8.3*). We don't stop— or must we? Try it for yourself. How many times can you fold the paper?

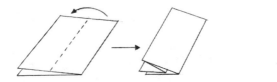

Fig. 8.3

As we saw in Chapter 5, the thickness of a sheet of paper, though modest, is not zero. If we could fold the paper forty-two times, it would reach from our desk to the moon. Fifty-one folds would produce a paper tower extending well past the sun. Plainly, forty-two folds is impossible. In fact, eight folds is impossible.

Our purpose here is not to set a new Olympic record for folding a sheet of paper. Instead, we'll simply fold it as many times as we are able (five or six times) and then unfold the sheet to see what we have created. We urge you to attempt these daring folds of repeated right-over-lefts and play along with us.

When we unfold the paper after we've made as many folds as we can, we see a crinkled mess completely devoid of structure. Once we get over the initial shock and awe of the crumpled chaos, we look closely and make a trivial observation: The folds come in two flavors—downward folds and upward folds. We will call the downward folds "valleys," represented by the symbol V, and refer to the spiky upward folds as "ridges," denoted by Λ (*Figure 8.4*). The mere act of naming objects often allows us to understand them better.

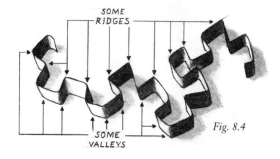

Fig. 8.4

If we read the folds of the folded paper shown in Figure 8.4 from left to right, we see the sequence that begins:

Valley Valley Ridge Valley Valley Ridge Ridge ...

Looking at this sequence of valleys and ridges, no discernible pattern jumps out at us. But why not? After all, we did our folding in a methodical, right-over-left way; we'd expect that the consistent repetitions would translate into *some* pattern rather than this complex mess. But when we're faced with complexity, a good strategy is to turn away and start over with something simple. In this case, the cause of complication is clear: so many damned folds! So let's start with a clean slate, or more accurately a clean sheet, and this time repress our yearning to fold uninhibitedly.

A FEW GOOD FOLDS

If we just fold once and then open the paper again (*Figure 8.5*), we see one lone valley—so far so good. If we fold twice (*Figure 8.6*), we see two valleys followed by one lone ridge. If we fold three times (*Figure 8.7*), the creases spell out:

Valley Valley Ridge Valley Valley Ridge Ridge

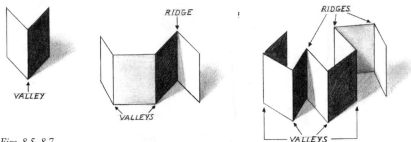

Figs. 8.5–8.7

If we now list these "paper-folding sequences" and look at them collectively, then we cannot help but notice a trendy surprise:

Valley

Valley Valley Ridge

Valley Valley Ridge Valley Valley Ridge Ridge

There's a pattern: The first part of each new sequence is exactly equal to the sequence above it. To confirm this observation, let's do one more fold: This time we fold four times before we unfold (*Figure 8.8*). We append the result at the end of our previous list, saving

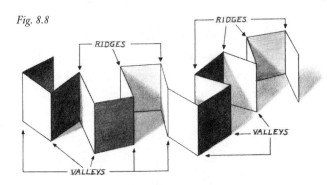

Fig. 8.8

space by using our symbols V and Λ instead of spelling out the words "valley" and "ridge":

V
VVΛ
VVΛVVΛΛ
VVΛVVΛΛVVΛΛVΛΛ

And sure enough, this surprising pattern continues if we do a fifth fold, and a sixth. We see that the first part of the five-fold sequence agrees with the four-fold sequence, and the first part of the six-fold sequence matches the five-fold sequence:

V
VVΛ
VVΛVVΛΛ
VVΛVVΛΛVVΛΛVΛΛ
VVΛVVΛΛVVΛΛVΛΛVVΛVVΛΛΛVVΛΛVΛΛ
VVΛVVΛΛVVΛΛVΛΛVVΛVVΛΛΛVVΛΛVΛΛVVΛVVΛΛVVΛΛVΛΛΛVVΛVVΛΛΛVVΛΛVΛΛ

This observation provides us with the insight to predict the first half of the paper-folding sequence for any given number of folds simply by knowing the sequence of folds for one less than that number. If we represent the folds with symbols and record the sequences, patterns come into focus.

Once we find structure, we cannot help looking for more. In this case, we begin by trying to predict the center term in the sequence. We've marked each center term below with an underline.

<u>V</u>
VV<u>Λ</u>
VVΛ<u>V</u>VΛΛ
VVΛVVΛΛ<u>V</u>VVΛΛVΛΛ
VVΛVVΛΛVVΛΛVΛΛ<u>V</u>VVΛVVΛΛΛVVΛΛVΛΛ
VVΛVVΛΛVVΛΛVΛΛVVΛVVΛΛΛVVΛΛVΛΛ<u>V</u>VVΛVVΛΛVVΛΛVΛΛΛVVΛVVΛΛΛVVΛΛVΛΛ

Again we see a surprising coincidence: Each center term is a V. Now how can we predict the second half of the sequences? The valleys and ridges in these sequences arose organically from the process of folding a piece of paper right over left. Thus it is natural to wonder what would happen if we folded one of the sequences of symbols in half through that center V—that is, if we rotated the right half of the sequence over the left half of the sequence as though it were hinged on that center V (*Figure 8.9*).

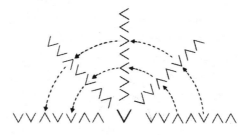

VVAVVAA V VVAAVAA   *Fig. 8.9*

Surprise. If we rotate the right half of the sequence onto the left half, then those two halves match up perfectly. For example, if we rotate the right half of the sequence VVA—that is A—then it becomes a V and so coincides with the left V (*Figure 8.10*). Similarly, if we fold VVAVVAA in half, then the right side VAA rotates over to become VVA and thus is identical to the left half of VVAVVAA (*Figure 8.11*). This fold-and-match symmetry continues throughout each folding sequence.

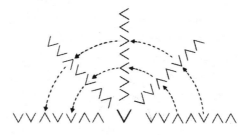

Figs. 8.10 and 8.11

Thus we are able to produce the paper-folding sequence for seven folds by writing down the sequence for six folds, appending an extra V (the soon-to-be-center V), and then rotating the six-fold sequence over the newly added V as though it were hinged on the V. We would see the impressively long run:

VVAVVAAVVVAAVAAVVVAVVAAAVVAAVVAVVAVVVAAVVAAVAAAVVVAVVAAVAAVVVAVVAAVAAVVAAVVAAVVAAVVAAVVVAVVAAVVAAVVAAVVVAAVAAVVAAVAAVVAAVVAAVVAAVVAAVAAVAA

CHAOS NO MORE

We have just tamed the chaos! We can now produce the paper-folding sequence for any number of folds—even for those folds that are physically impossible. Thus our discovery allows us to actually transcend the constraints of our physical world and our physical paper. Even though we can physically fold a paper only seven times, we now see how to generate the correct pattern of folds for a paper theoretically folded eight times. We simply write down the seven-fold sequence, append a V̲, and then hinge the seven-fold sequence over that center V̲ to produce the second half of the eight-fold sequence. In theory we could continue the process up to and beyond the 51-fold sequence that would take us past the sun—but of course the string of V's and Λ's would itself extend far beyond the sun.

All these new insights followed from simply naming what we saw—the valleys and ridges—and searching for a pattern. Suddenly the creased chaos has become crystal-clear.

FROM SHEER CHAOS TO COMPLETE REGULARITY

And now for something completely different. If we explore the paper-folding sequences from another angle, we suddenly discover that there is a simple pattern hidden within the folds that will allow us to generate new sequences in a totally new way. The first step is to write the paper-folding sequences center-justified (with the center V̲ in the middle of the line). If we then focus on every other term, starting with the first (these are in boldface below), we see something quite unexpected:

V

VVΛ

VVΛVVΛΛ

VVΛVVΛΛVVVΛΛVΛΛ

VVΛVVΛΛVVVΛΛVΛΛVVVΛVVΛΛΛVVΛΛVΛΛ

VVΛVVΛΛVVVΛΛVΛΛVVVΛVVΛΛΛVVΛΛVΛΛVVVΛVVΛΛVVVΛΛVΛΛΛVVΛVVΛΛΛVVΛΛVΛΛ

The boldfaced terms exhibit supreme regularity. In fact, they simply alternate—valley, ridge, valley, ridge, and so forth:

Ɐ∧

If we now focus on the non-boldfaced terms, something truly remarkable comes into focus. For example, let's consider the four-fold sequence:

ⱯⱯ∧ⱯⱯ∧∧ⱯⱯⱯ∧∧Ɐ∧∧.

We now remove the boldfaced, alternating terms to reveal

Ɐ Ɐ ∧ Ɐ Ɐ ∧ ∧ ,

a sequence with a familiar feel. Do you recognize it? It is exactly the three-fold sequence! If we check a few other sequences we'll see that the same thing happens: What remains after we remove the alternating terms from any paper-folding sequence is precisely the previous paper-folding sequence. Thus we see a surprising, somewhat incestuous behavior among these sequences of valleys and ridges.

We can use this fantastic structure to generate new paper-folding sequences by inserting terms. Given any paper-folding sequence, we can build the next sequence by simply taking the one we have, placing spaces between adjacent terms, and then weaving in a run of alternating valleys and ridges. For example, given the three-fold sequence ⱯⱯ∧ⱯⱯ∧∧, we can build the four-fold sequence by separating the terms and weaving the down-up sequence into the mix:

ⱯⱯ∧ⱯⱯ∧∧	(The three-fold sequence.)
Ɐ Ɐ ∧ Ɐ Ɐ ∧ ∧	(The three-fold sequence spaced out.)
ⱯⱯ∧ⱯⱯ∧∧ⱯⱯⱯ∧∧Ɐ∧∧	(Weaving in the alternating sequence, the result is the four-fold sequence!)

So we have uncovered two different hidden patterns that allow us to generate the paper-folding sequence in two different ways. Our first method to generate the next folding sequence is to begin with the previous sequence, append a V, and then rotate a copy of the sequence over the V. The second way of producing the same sequence is to start with the previous folding sequence, spread it out, and weave the simple down-up sequence (valley, ridge, valley, ridge) into the spaces.

Although we've tamed what initially seemed to be chaos, both generating processes suffer from what appears to be a necessary evil: They both require that we know the previous paper-folding sequence in order to determine the next one. Obviously, to fold the paper, say, six times, we must first fold it five times, so somehow it seems natural that the five-fold sequence must enter into the fold as we move to the six-fold sequence. Natural, sure; but is it necessary? The answer to this question provides a wonderful opportunity to consider ideas that are the cornerstones to the present computer age.

TURNING TO TURING

Behind every computer, every ATM, every Web transaction, every computer worm, and every spam message asking us to help some stranger legally wire $2,000,000,000,000 into an American bank, there is a computer program hard at work. The programming that is now so prominent in so much of our lives is, in fact, a relatively new art. The architect of the creative art of computer programming was the great British mathematician and computer scientist Alan Turing, who first introduced the concept of a computer in 1932. His idea, known as the *Universal Turing Machine*, led to the building of the first digital computers in 1950.

Turing is known for many important scientific contributions, including the brilliant insights that made possible the deciphering of the *enigma code*, the code used by the German armed forces in World War II in radio communications. Here we will explore a version of Turing Machines known as *finite-state automata*. Roughly speaking, finite-state automata are the simplest computers we can imagine— far simpler than your laptop, but with surprisingly far-reaching capabilities.

Finite-state automata are very simple machines: All they do is read and write numbers according to certain rules, which collectively are better known as a *program*. Envision a ticker tape of numbers running through a machine with a reader and a writer. (We don't mean a *human* reader or writer, of course.) The ticker tape of numbers passes through the reader, which reads each number individually. After the reader reads a number, the writer writes some numbers at the end of the ticker tape, and then the reader slides over and reads the next number on the list. The numbers written at the end of the list are based on the number read and are determined by the specific program. Figure 8.12 shows the general idea.

Fig. 8.12

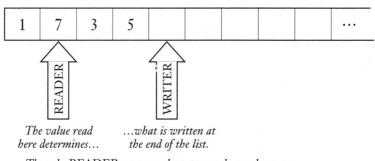

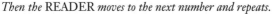

The value read   *…what is written at*
here determines…   *the end of the list.*

Then the READER *moves to the next number and repeats.*

To make the notion of finite-state automata concrete, let's consider a specific example. In this "program," the set of rules is essentially arbitrary—there is no method to the substitution madness:

If the reader **reads** a 1, then the **writer** writes 3, 2 at the end of the list.
If the reader **reads** a 2, then the **writer** writes 0, 2 at the end of the list.
If the reader **reads** a 3, then the **writer** writes 3, 1 at the end of the list.
If the reader **reads** a 4, then the **writer** writes 4, 1 at the end of the list.
Then the reader slides over and reads the next number on the list.

To run this program, we need at least one number on our ticker tape. Suppose we start with 1. Since the reader reads that 1, the writer writes 3, 2 at the end of the list. Now the reader slides over and reads the next number, which is the new 3. Having read that 3, the computer writes 3, 1 at the end of the list and then reads the next number, 2. The output looks like this:

$$1, 3, 2, 3, 1, 0, 2, 3, 1, 3, 2.$$

The computer is not happy when it reads that 0, because it has been given no instruction for what to write when it reads a 0. So what does it do? It stops. When a program stops, computer scientists say the program *halts*. Halting may mean that the program has gracefully completed its mission, but in many cases it is a euphemism for crashing. Our example is not that complicated, so we can see the flaw in the program right away, but in general it is very difficult to determine if a Turing Machine halts after a finite number of steps or not. This important computer science issue is known as the *halting problem*. Scientists can prove that it is impossible to devise an ironclad general method for determining whether a program will halt or not.

As it turns out, this simple reading and writing procedure can be used to create computer programs that perform extremely complicated calculations. In fact, all computer programs, no matter how complicated, from word processors to Internet search engines to electronic banking, are ultimately based on the simple reading and writing rules of this abstract Turing Machine. Of course, the challenge is to create the rules that will perform a desired task.

Instead of fretting over issues that keep computer programmers up into the wee hours of the morning, let's just consider another example of finite-state automata. In fact, we'll consider the previous program with one small change—we'll get rid of that pesky 0. Here is the new set of rules for our machine:

If the reader **reads** a 1, then the writer **writes** 3, 2 at the end of the list.
If the reader **reads** a 2, then the writer **writes** 4, 2 at the end of the list.
If the reader **reads** a 3, then the writer **writes** 3, 1 at the end of the list.
If the reader **reads** a 4, then the writer **writes** 4, 1 at the end of the list.
Then the reader slides over and reads the next number on the list.

If we start this program with a ticker tape containing 1, then it is easy to see that this program will never halt, since every value that is written by the writer (1, 2, 3, or 4) is an allowable value that can be read by the reader. So this program will run forever and spit out an endless list of digits. Given 1 as the starting number, the computer puts out a string that begins like this:

$$1, 3, 2, 3, 1, 4, 2, 3, 1, 3, 2, 4, 1, 4, 2, 3, 1, \ldots$$

Perhaps we should not be impressed: It appears that this simple program is generating an uninteresting random list of infinitely many numbers. But before we move on, let's simplify the list by noting whether the numbers are odd or even. Let's replace each odd number by an odd symbol—say, V—and replace each even number with an even odder symbol—say, Λ. When we replace the output digits with these symbols, we produce:

VVΛVVΛΛVVVΛΛVΛΛVVVΛVVΛΛΛVVΛΛVΛVVVΛVVΛΛVVVΛΛVΛΛΛVΛVVΛΛVVΛΛVΛΛ. . .

In an incredible turn of events, it turns out that we are generating the paper-folding sequence for arbitrarily many folds! Thus the paper-folding sequence is in actuality the output of an extremely simple, five-line Turing Machine program. Perhaps more surprising, we no longer require the previous folding sequence to generate the next one. Using the five rules and the "starting seed" 1, we are able to move right through those creases to produce the paper-folding sequence for arbitrarily many folds.

The early terms in the sequence organically generate the future terms—all the way past the sun, in the case of fifty-one folds. Just those five simple rules, amazingly, set a perfect folding course for as far as we wish to travel. Our up-and-down movement as we undulate along the paper, in fact, holds within it the whole future story of our folding future—miles and miles and miles away at a horizon we'll never reach. There is indeed much more structure to this chaotic-looking, folding jumble than first met our eye.

FROM FOLDED SWANS TO FOLDED DRAGONS

After all these pages we confess that we have made zero progress in understanding how to bring an origami swan to life. Hoping to duck the fowl issue, here we move beyond the delicate and petite swan and ignite a process for folding paper into a fiery dragon. Our dragon tale has both good news and bad news. On the bright side, our method requires no origami skills whatsoever—we will simply employ the trivial right-over-left fold that has become the mainstay of this entire discussion. As we will see, however, there will be a dark side.

Our paper-folding sequences arose from folding right over left some number of times and then carefully unfolding the paper. We read off the valleys and ridges created by the folds, and we generated our sequence. We now return to the creased paper itself and ask, What shape do we see if we arrange the paper so that each fold forms a right angle of 90 degrees? For example, with one fold we see a 90-degree angle (*Figure 8.13*). Adjusting the paper after two folds so that all angles are right reveals a "saucepan"-esque image (*Figure 8.14*).

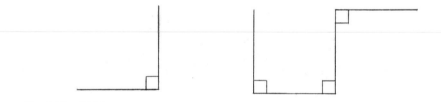

Figs. 8.13 and 8.14

Starting with the two-fold image, can we predict the three-fold image? The answer is yes—and we already know how. We recall that one method of generating the next paper-folding sequence is to undulate through the current sequence, weaving in the valley-ridge-valley-ridge sequence. Here we can use this process visually by inserting a sequence of alternating over-under folds between the creases of our right-angled arrangement (*Figure 8.15*).

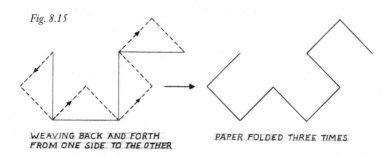

Fig. 8.15

WEAVING BACK AND FORTH
FROM ONE SIDE TO THE OTHER

PAPER FOLDED THREE TIMES

The next fold reveals an even more exotic form (*Figure 8.16*). This image may be familiar to careful readers of Michael Crichton's

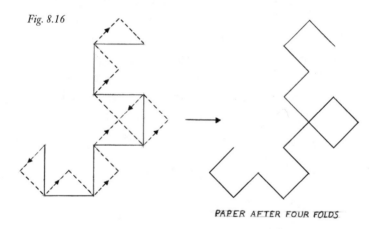

Fig. 8.16

PAPER AFTER FOUR FOLDS

1990 novel, *Jurassic Park*. The story is divided into sections called "Iterations," each of which opens with a quote from the fictional character Ian Malcolm and an image that is a metaphor for the increasing complexity of the plot line. The First Iteration includes the image in our Figure 8.16 together with the quote, "At the earliest drawings of the fractal curve, few clues to the underlying mathematical structure will be seen." We now realize that this icon is nothing more than one of the steps in our paper-folding process, unfolded and arranged at right angles—a highly appropriate image, since we can view the steps in paper folding as iterations of a simple process.

Crichton's Second Iteration contains an image of the paper-folding sequence with one additional fold, again in its right-angle configuration (*Figure 8.17*). Here Ian professes, "With subsequent drawings of the fractal curve, sudden changes may appear." The

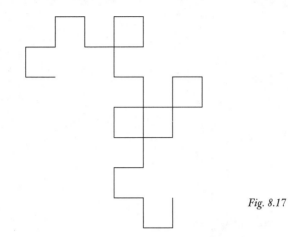

Fig. 8.17

Third Iteration (*Figure 8.18*), the next iteration of the paper-folding curve, contains a number of squares; these are instances where the paper corners meet. "Details emerge more clearly as the curve is

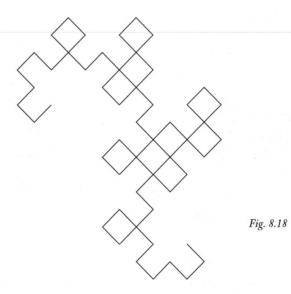

Fig. 8.18

redrawn," Ian exclaims. As *Jurassic Park* progresses, the images become increasingly complex and Ian's pronouncements increasingly dark: "Inevitably, underlying instabilities begin to appear" (the Fourth Iteration, *Figure 8.19*); "Flaws in the system will now become severe" (the Fifth, *Figure 8.20*); "System recovery may prove impos-

Figs. 8.19 and 8.20

sible" (the Sixth, *Figure 8.21*); and finally at the Seventh Iteration (*Figure 8.22*), "Increasingly, the mathematics will demand the courage to face its implications."

Certainly there is no doubt in our minds: The paper-folding

Figs. 8.21 and 8.22

image that is coming into focus is none other than the fiery dragon that we first encountered at the opening of this chapter. At that initial confrontation, we assumed that the process required to generate such an infinitely complicated object seemed far beyond our mathematical abilities. However, after our journey through the folds, we now see that our fire-breathing foe is actually the result of an extremely simple process—folding paper right over left.

The *Dragon Curve* is the end result of infinitely many folds. Thus we see the dark side of the Dragon Curve—to produce a perfect one, we must perform the trivial folding task an *infinite* number of times. Of course the Turing Machine we built provides us with the endless instructions, but actually *performing* those endless moves would take more than a lifetime of effort. On the bright side, we see that what first appeared incomprehensible can be made crystal-clear by simply searching for pattern and structure.

As you may have suspected when you read some of Ian Malcolm's pronouncements, the Dragon Curve is an example of a *fractal*. A fractal is any geometric object that has infinite complexity. Quite often fractals have a self-similar nature in which a magnified portion resembles the entire object (*Figure 8.23*). In many cases, a fractal

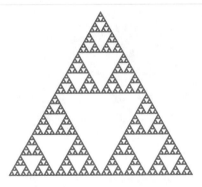

Fig. 8.23

image is created by a simple process that is repeated over and over again. Thus the Dragon Curve beautifully captures the fractal spirit.

WALL-TO-WALL DRAGONS

This wild intellectual journey has taken us from rudimentary paper folding, to finding patterns, to simple computer programs, to fiery dragons. We end this journey with a final, fitting surprise. Since the Dragon Curve is created by folding paper infinitely many times, it contains an infinitely jagged boundary. That crinkled skin is made up of infinitely many creases. What is totally unexpected is that the dragon's infinitely crinkled skin has a narcissistic self-appeal. That is, if we duplicate the Dragon Curve several times and view the resulting copies as jigsaw puzzle pieces, then those infinitely jagged identical pieces snap together perfectly to cover the entire plane (*Figure 8.24*)!

Fig. 8.24

The realization that the skin of the Dragon Curve fits perfectly together with other copies of itself is a further testament to the seemingly endless amount of structure that the simple paper-folding process possesses. It also offers another tiling of the plane in which we require only one type of tile. In Chapter 7 we saw a method of tiling the plane using Golden Triangles. The Golden Triangle tiling and the Dragon Curve tiling create a conceptual yin and yang of simplicity and complexity. The Golden Triangles are very simple tiles but can be placed on the plane so that the pattern they create is chaotic—the pattern is infinitely complicated and never repeats. Here, we found Dragon Curve tiles that are infinitely complicated but fit together in a periodic manner—that is, in a manner that repeats at regular intervals. In tiling your bathroom, you are now flush with philosophical possibilities that may be far too draining. (Sorry.)

Through the simple folds of a piece of paper, we have discovered beautiful patterns, explored the birth of modern computing, and tamed the infinite complexity of Dragon Curve fractals. Even more important, we have seen the power of searching for structure. That quest for structure is in actuality a journey toward understanding. It's a journey that takes us first to distant worlds but then, with twists and turns, crimps and wrinkles, leads us back to our own world. In the next chapter we take twisting and turning to their ultimate extremes and discover worlds that are at once foreign and familiar.

CHAPTER 9

A TWISTED TURN IN AN AMORPHOUS UNIVERSE

An Exploration of an Elasticized World

The moving power of mathematical invention is not reasoning but imagination.—Augustus de Morgan

Obviously . . . Tie your ankles together with a rope five feet long. Shuffle into the privacy of your bedroom. Now, without removing the rope, drop your pants and then attempt to put them back on *inside out*. Of course such a feat must be physically impossible, defying even the skills of the Great Houdini.

Surprise . . . This trouser-inversion trick is absolutely possible. Your slacks can be pulled, pushed, and contorted until they finally fit neatly over your lower limbs with the zipper proudly in front and the pockets waving outside by your hips. This frivolous exercise is tailor-made to foreshadow the surprising and elastic possibilities within a rubber-made world.

A WAY TO WONDER

We often hear that we should be flexible. While some may take this advice to heart and take up yoga, here our goal is to develop agility in our thinking. In this chapter we take flexibility to the ultimate elastic extreme, exploring a universe in which everything we touch and behold is perfectly amorphous.

One method we can all use to pour forth one creative idea after another is to begin with our everyday world, imagine some subtle property slightly altered, and then explore this altered state. Which features remain the same? Which features are different? Exploration of a hypothetical world generates whole galaxies of new ideas. And we'll discover a synergistic interplay in which those new ideas will lead us back to new insights into our familiar, everyday world.

FUN WITH RUBBER—EXOTIC ADVENTURES IN A RUBBER-SHEET WORLD

Here we take our own advice and imagine our universe with one small change: We imagine that every single part of our physical world is wildly distortable. That is, every object is made of an unrealistically flexible, rubbery substance that is elastic without bound. Each object can thus be stretched, bent, compressed, expanded, and generally morphed at will. In this realm, a basketball can be inflated to the size of the moon and a conductor's baton can be stretched and bent to resemble the Saint Louis Gateway Arch. In this world a dollar bill can be stretched as far as we wish, which, given today's gasoline prices, sounds extremely attractive. This seemingly silly fantasy of elasticized reality belongs to a mathematical area known as *topology* or, more informally, "rubber-sheet geometry."

Of course, this hypothetical domain would have no interesting features without some constraints—if everything could simply dissolve into a structureless soup, there would be no distinctions left to contemplate. Instead, we imagine that objects in this world are made of molecules that hang on to neighboring molecules, and although the molecules can be stretched, shrunk, and twisted, the bonds are not to be broken, and new bonds are not to be introduced. Thus objects can be stretched and contorted, but they can be neither cut nor glued. For if an object were torn, then some bond would be broken; and if, for example, we created a circle by gluing the two ends of a line segment together, then we would be creating a bond that did not exist before (*Figure 9.1*).

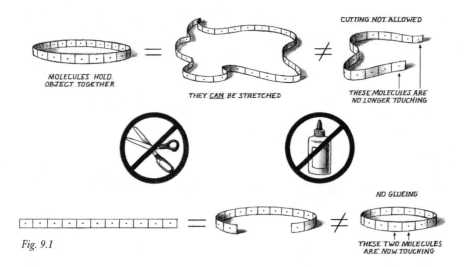

Fig. 9.1

To develop our intuition into what can and cannot be legally realized in this rubbery realm, we warm up with some flexible fonts. What letters of the alphabet can be morphed to resemble one another? For this exercise the font makes a difference, and we will choose a sans serif font (without the little horizontal spurs). Thus **S** can be distorted to look like **C, I, J, L, M, N, U, V, W,** and **Z**. All these letters are merely distortions of a line—we can create each of those curves, bends, and angles by contorting that line. Likewise, **O** is the same as **D**. Another group of letters consists of **E, F, T,** and **Y**; each of those can be bent and stretched to look like any other in the group (*Figure 9.2*). By looking at the world in this new manner, we are asking which objects or figures can be stretched, shrunk, bent, or other-

Fig. 9.2

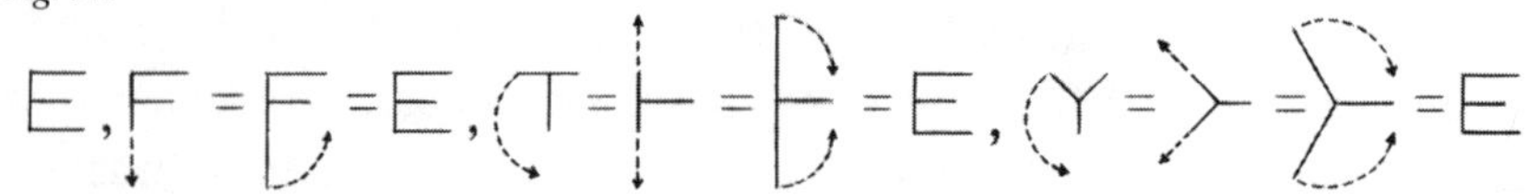

$$E, F = F = E, (T = \vdash = \vdash = E, (Y => \rangle - = \Sigma = E$$

wise distorted to resemble one another and which cannot. The letter O cannot be stretched to look like the letter X, for example, because the X has a point where four lines emerge, whereas every point on the circle has just two lines emanating out (*Figure 9.3*). No matter how we stretch the X, the point with four lines coming out will always have four (possibly curvy script) lines emerging.

Fig. 9.3

Looking at the world in this distorted topological way is at least mildly entertaining, but we may not be convinced that this elastic perspective is of any relevance to our everyday lives. Here we silence that concern by tackling the challenge of those annoying metal tavern puzzles.

TAVERN PUZZLES

So-called tavern puzzles, made of metal with a forged-by-a-blacksmith appearance, are incredibly frustrating. We're supposed to remove an attached ring, push a wooden ball on a chain through a hole that is obviously too small, or rearrange other elements of the puzzle in seemingly impossible ways. We pick up one of these puzzles at our peril—if we can't solve it, we're flushed from the embarrassment as well as the beer—a sobering thought.

In fact, those irksome tavern puzzles can be solved, but they require a sequence of very ingenious moves (the first being to move one's beer away from the action). But what if we were topologists and those puzzles were made of stretchable rubber rather than unforgiving metal? Then solving them would be a simple matter. Interestingly enough, the exercise of solving a rubber version of the puzzle can sometimes lead us to a method for manipulating the unbending metal to achieve the same goal. Let's consider an example.

In order to solve the puzzle pictured in Figure 9.4, we are to remove the heart-shaped ring. How would we proceed if the puzzle

Fig. 9.4

were made of rubber rather than metal? We would first simply shrink the metal bar to morph the puzzle along the lines shown in Figure 9.5. The solution is now easy; without any further contort-

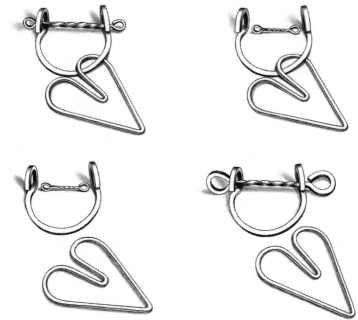

Fig. 9.5

ing, we see how to remove the ring. Here's the twist: We now consider making the same sequence of moves (except the shrinking) on the original, undistorted metal puzzle, and we ask ourselves whether we can have the ring follow the steps that we get by undistorting our distorted solution. We see that we can (*Figure 9.6*), and thus realize that those tavern puzzles become a piece of cake if we are flexible in our thinking. (We must confess, though, that we find *some* of these puzzles almost impossible to solve even in the sober light of day and armed with our flexible thoughts.)

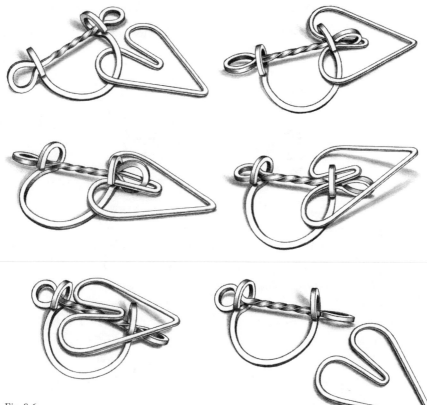

Fig. 9.6

SARTORIAL SHENANIGANS—RUBBER UNDIES AND DROPPING TROU

By applying this rubber thinking to our wardrobes, we find provocative alternatives to our conventional methods for disrobing. We begin with the following question: Is it possible to remove a pair of sufficiently stretchable underwear without removing one's pants?

Revealing a brief answer may appear to be in mildly poor taste to some genteel readers; however, in the name of mathematics we must face challenges wherever they may expose themselves. The more daring and less modest readers might enjoy the topological challenge of working out the answer for themselves before reading on.

It is indeed possible to remove rubber undies in this bizarre manner, and the tavern puzzle method can be employed to see how. First we suppose that our left leg is made of rubber and can be dramatically compressed to a length of two inches (*Figure 9.7*). Then a

Fig. 9.7

non-stretchable pair of underwear can easily be pulled around the foot of our new short left leg and then made to fall down the right pants leg around our right limb (*Figure 9.8*). With this (possibly disturbing) rubber-left-leg solution in our minds, we can now envision how to perform the step of moving the underwear past the left leg when it is its usual full length, because now the underwear is stretchable.

Fig. 9.8

Discovering this strategy of rubber underwear removal illustrates how by focusing on the essential features of an issue—whether they are topological, personal, or even political—and forgoing the superficial ones, we are led to new insights and breakthroughs that would otherwise have remained out of sight. On the other hand, keeping our underwear out of sight does have a certain appeal.

At the opening of the chapter we posed the challenge of tying our ankles together with a five-foot-long rope and, without untying the rope, taking off our pants and putting them back on *inside out*. This wardrobe inversion is completely possible with actual pants. No assumption of unrealistic elasticity is required—happily, we can leave our pair of Haggar stretchable slacks in the closet. The best strategy for unraveling this puzzle is just to try it, although it's far from easy. We do recommend that you enjoy the struggle before sneaking a peek at Figure 9.9, which shows all the right moves.

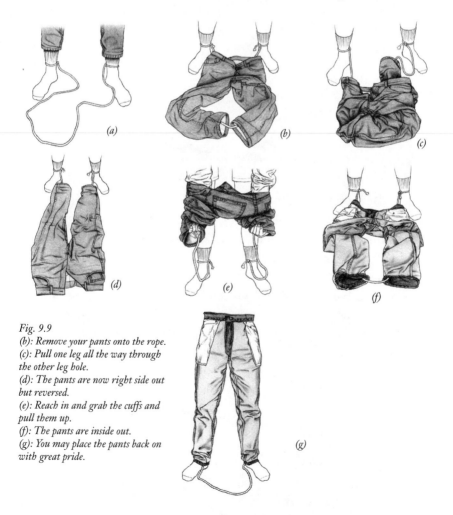

Fig. 9.9
(b): *Remove your pants onto the rope.*
(c): *Pull one leg all the way through the other leg hole.*
(d): *The pants are now right side out but reversed.*
(e): *Reach in and grab the cuffs and pull them up.*
(f): *The pants are inside out.*
(g): *You may place the pants back on with great pride.*

IS EARTH ACTUALLY THE SHAPE OF A DOUGHNUT?

Let's extend our field of view from the narrow confines of our dressing rooms to the expansive majesty of our entire world. We all know that the Earth is shaped like a ball. But if our world were enormously elastic, we would have a different intuitive notion of which shapes are the same. A ball-shaped world and a banana-shaped world would be equivalent, since a ball can be stretched and distorted to resemble a banana (*Figure 9.10*). But even a fabulously flexible elastic world would have some constraints and limitations. For example, as we will show mathematically, a doughnut cannot be stretched and distorted into the shape of a ball. Even in a rubber-sheet solar system, Earth differs from a glazed doughnut.

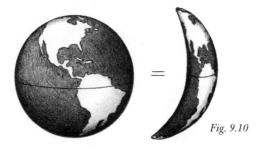

Fig. 9.10

We know it intuitively, but how do we prove mathematically that even in a rubber-sheet universe it's impossible to stretch and distort a doughnut to make it resemble a ball? Actually, we've all experienced some relationship between these objects in our everyday lives: After consuming the former *we* tend to more closely resemble the latter. But biting into a doughnut and chewing it are not permitted in this calorie-conscious study of rubberized geometry (remember, breaking and cutting are not allowed on the topology diet).

As we've seen, elastic properties can lead to surprising outcomes involving inverted garments. So without exploring further, we can't be sure that it really is impossible to stretch a doughnut into a ball. We need to identify a feature of the doughnut-shaped object that, first, is preserved during any allowable distortion and, second, is a feature that the ball does not possess.

One feature that meets the first requirement comes into focus when we draw a circle along the surface of a doughnut in such a way that the circle passes through the doughnut's hole (*Figure 9.11*). We observe that this circle does not separate the surface of the doughnut into two pieces; that is, we can journey along the surface of the doughnut from any point to any other point without crossing that loop (*Figure 9.12*). In fact, if we were to reshape the doughnut in any

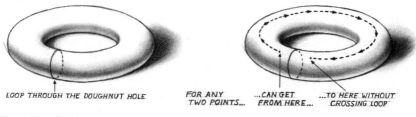

LOOP THROUGH THE DOUGHNUT HOLE   FOR ANY   ...CAN GET   ...TO HERE WITHOUT
                                  TWO POINTS... FROM HERE... CROSSING LOOP

Figs. 9.11 and 9.12

manner not involving cutting or gluing, then we would see that our loop, while it might be wildly wiggly, owing to the distortion, would still not separate any two points on the surface of the distorted doughnut from each other (*Figure 9.13*).

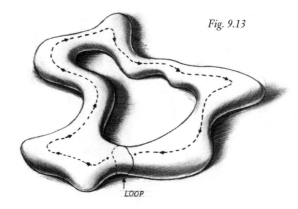

Fig. 9.13

LOOP

So what happens when we draw a circle on the surface of a ball? It turns out that no matter how we draw it, that circle will always separate the ball's surface into two regions. That is, there will always be pairs of points on the sphere that are separated by the newly drawn circle (*Figure 9.14*). This property does not change if we distort (without cutting or gluing) the sphere (*Figure 9.15*). No matter how we twist and shout, any curved loop will partition the sphere into two separate pieces.

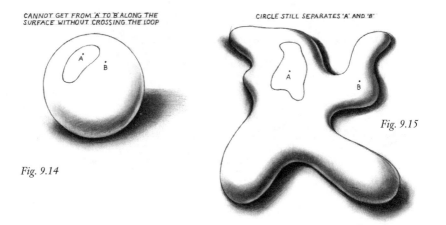

CANNOT GET FROM 'A' TO 'B' ALONG THE
SURFACE WITHOUT CROSSING THE LOOP

CIRCLE STILL SEPARATES 'A' AND 'B'

Fig. 9.15

Fig. 9.14

But let's look at this issue from another angle, just to make sure we've got it right. Suppose for a moment that we *could* reshape the doughnut into a ball. Then our circle looping through the doughnut's hole would be stretched and bent during the distortion and would end up being a curvy loop on the surface of the sphere (*Figure 9.16*). Thus the one-piece surface of the doughnut (without our cir-

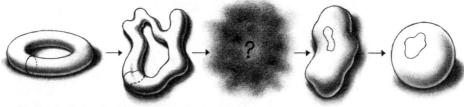

Fig. 9.16   *An imagined distortion.*

cle) would be distorted into two separate pieces on the surface of the ball (*Figure 9.17*). In other words, by merely stretching and shrinking, but not cutting, we were able to transform a one-piece surface into a two-piece surface, which is impossible: The only way to split a single piece into two pieces is to make a cut.

So we have rigorously confirmed our intuition that, even in a

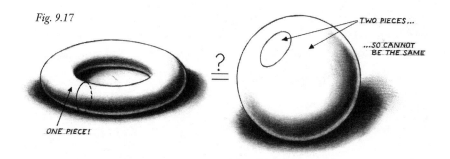

Fig. 9.17

TWO PIECES...

...SO CANNOT BE THE SAME

ONE PIECE!

world of unreasonable distortability, a ball is different from a doughnut. In this happy circumstance we see that mathematical thinking has confirmed our intuition. Unfortunately, when we propose to stretch our minds around a diamond ring, the marriage between reality and our intuition leads to an uncomfortable separation.

TWO HOLES, ONE RING, ZERO DIVORCES

Suppose that we come across a rubber disk with two holes and a diamond ring looped through those holes (*Figure 9.18*). It appears intuitively obvious that we could not stretch and contort the rubber disk so that the ring would loop through only one hole (*Figure 9.19*). To

Figs. 9.18 and 9.19

achieve such a result, we would have to cut the rubber disk and then glue it back together (*Figure 9.20*)—it would be impossible to free the ring from one of the holes by simply stretching.

Fig. 9.20

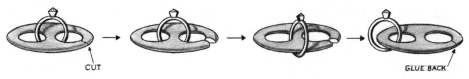

CUT

GLUE BACK

Surprise. Such a seemingly impossible act is in fact possible. Recalling that a picture is worth a thousand words, we present the details of this fantastic topological feat in Figure 9.21.

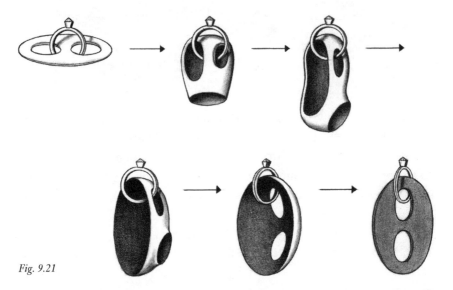

Fig. 9.21

As you can see, the first step is small and straightforward—we simply let the rubber disk go limp and stretch one hole to make it somewhat larger. The next step is also an easy stretch—we pull the edge of the now-larger hole over the ring. The move that follows illustrates that we are stretching without cutting; there are no tricks, mirrors, or sleight of hand. Recall that we are working with elastic of unlimited flexibility, and it is up to us to keep our minds equally flexible. Continuing we soon see that the ring that was clearly looped through two holes now appears to be looped through only one.

Even after we've studied the images, one mystery remains: How was it possible for us to remove the ring, without cutting, from one hole? The surprising answer is that we didn't; after our moves, the ring still passes through the same two holes! If we follow the stretching closely, we see that what was the small hole on the right is now stretched out to be the outside curve of the rubber sheet, and what was originally the outside perimeter of the rubber sheet has become the small, unlinked hole on the right (*Figure 9.22*). The ring is still

Fig. 9.22

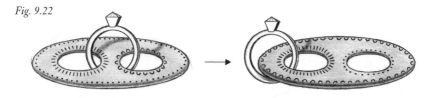

linked around both "holes"—we merely turned one of those holes into the outside boundary.

This ring-removal illusion defies our intuition and illustrates the wonders and strange twists possible within an amorphous universe. But more important, the puzzle provides an insight into what has been a recurring theme throughout the book—how we think and how we make sense of the world. Here we realize what a *surprise* truly is—a moment in which we discover that our intuition runs counter to reality.

When we are surprised by a particular outcome or event, we should consciously acknowledge that there must be a gap between our perception and reality. A surprise should be a signal inviting us to realign our intuition and our thinking so that they conform to actuality. One of the life lessons that mathematical thinking offers us is that we should always reexamine a surprising situation from various angles and points of view until that surprising feeling is replaced by a rock-solid intuitive understanding of the truth.

Our next eye-opening surprise centers around the following topological challenge: Is it possible to distort a linked two-holed "doughnut" (*Figure 9.23a*) into an unlinked two-holed "doughnut" (*Figure 9.23b*) without cutting or gluing? The eye-opening, unexpected answer is . . . well, see for yourself. Without detailed commentary, we present here a series of pictures, with each picture differing from the next by a small, easily doable stretch (*Figure 9.24*).

Fig. 9.23a

Fig. 9.23b

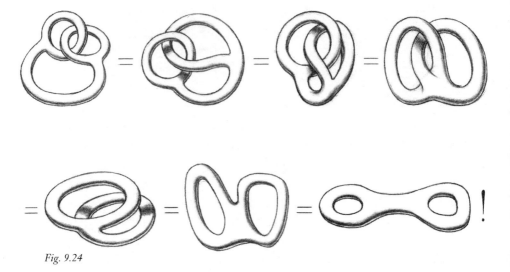

Fig. 9.24

As frivolous as these puzzles may seem, such topological constructs do in fact have real-world ramifications. We now journey from the surprises within undressing and unlinking to the twisted microscopic universe of life itself. Here the surprise is how our ever-twisting telephone cords can lead to insights about the secret world of DNA.

UNTANGLING SNARLED CORDS AND SNARLED DNA

Long before the age of cell phones we all had corded phones that would mysteriously tangle up (*Figure 9.25*). And even before the

Fig. 9.25

corded phone, we all had DNA that mysteriously tangles up within the nucleus of each of our cells (*all* our cells, not just the ones we hold up to our ears in airports). The corded telephone gave Ma Bell a communications monopoly (until she was forced to give birth to all those baby Bells), while DNA gives each of us a monopoly on our individuality. How are these different worlds connected? The connection is made through the topological notion of knots.

A mathematical *knot* is simply a closed loop of string that may or may not be knotted. The simplest knot is a loop that contains no knot at all and is called the *unknot* (*Figure 9.26*). (The fact that the math community refers to the unknot as a knot is reason #73 why people tend to avoid socializing with mathematicians.) Normal people consider a knot to be a loop that is genuinely knotted. But what does it mean to mathematicians for a loop to be knotted (*Figure 9.27*)? Well, it means that the loop cannot be transformed into a round circle—a.k.a. the unknot—without being cut.

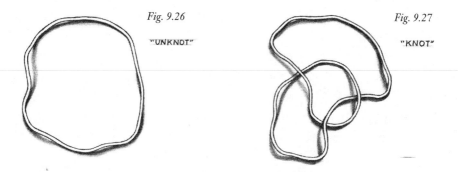

Fig. 9.26

"UNKNOT"

Fig. 9.27

"KNOT"

Unfortunately, it is not always easy to determine if a knotted tangle is really knotted or not. For example, can we figure out, just by looking, which of the tangles shown in Figure 9.28 can be untangled

Fig. 9.28

into the unknot? We think knot. As it happens, the jumble in the middle can be untangled into an unknotted circle, while the two flanking loops can never be unknotted completely. We should not be embarrassed, however, by our inability to determine genuine knottiness at first glance: No one knows a simple way to look at pictures of two knots and tell whether one knot can be manipulated to look like the other.

You might think it should be possible to study the patterns generated by the woven webs of overcrossings and undercrossings to determine which jumbles are truly knots, but no one has yet discovered a simple procedure that works for all jumbles. Of course, if we are allowed to cheat—that is, change various crossings by cutting and regluing the threads—then we can always unsnarl any knot to become the unknot. As we'll now discover, topology—the study of rubber-sheet geometry—brings us to the insight that life itself does not always abide by all the rules.

THE SPICE OF LIFE

The coiled double helix (*Figure 9.29*) is one of the most famous shapes in biology because it describes the structure of the molecule of life, DNA (deoxyribonucleic acid). Biologists report that who we are, the essence of our being, is encoded in the three-foot-long strand of DNA that resides in the nucleus of each and every cell in the human body. But it doesn't require a degree in biology to see that the microscopic nucleus of a cell is not a realistic place to store an object the shape and size of a yardstick. To stuff that long DNA strand into that tiny one-room nucleus, we really have to pack it in there. But compressing that DNA by squashing it into a single point would generate far too much potential energy for its own (and our)

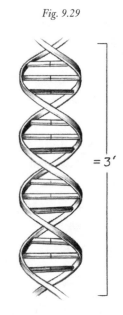

Fig. 9.29

= 3′

good. Thus Mother Nature takes a lesson from Ma Bell and the topology of twisted phone cords.

The coiled cord of a corded phone, shown in Figure 9.25, possesses some of the geometric properties of DNA. One of the most important is that the cord easily becomes twisted (and writhed as well) and thus compacted. We untangle these problematic twists by picking up the receiver and unwinding it until the cord returns to its normal coiled but untwisted shape. This simple exercise with corded phones illustrates nature's solution to the problem of storing DNA. DNA winds itself up like the phone cord in what is referred to as *supercoiling*. This allows the DNA to fit into tight spaces (such as the nucleus of your favorite cell) without building up excess energy. The next time you encounter a very twisted phone cord, don't be so annoyed. It's in that shape that nature found a topological way to have our DNA fit snuggly into a cramped nucleus. And it works perfectly—until the DNA has the urge to reproduce.

When a cell splits up, the two new cells don't want joint custody of that one DNA strand; each cell needs its own. So the DNA splits into two—one moves in with one cell, the other with the other cell. In theory we could imagine the beautifully spiraled DNA slowly untwisting and splitting rung by rung right down the center (*Figure 9.30*). Topology tells us, however, that this is impossible. The two

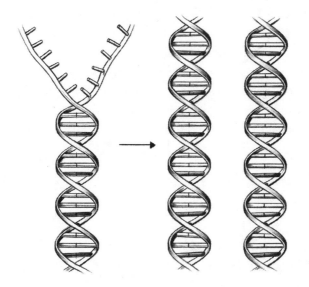

Fig. 9.30

sides of the DNA ladder are tightly coiled around each other, and then the DNA is supercoiled. In this gnarled configuration, it is impossible to pull the two strands apart after the rungs are split.

To convince yourself that separating the two coils is impossible, try bunching, twisting, and kneading together two three-foot-long pieces of string into a tight ball. Then try to separate the strings by grabbing two of the ends and pulling with great force. As you'll see, the strands are certain to be snarled. This simple experiment suggests that the image of the sides of a DNA ladder being pulled apart cannot be correct. Topological facts about linked strings irrefutably establish a biological fact about the replication of DNA without our needing to spend millions of dollars in laboratory experiments —here we see yet another illustration of the power of abstract mathematics.

During the separation, the two sides of the DNA ladder cannot remain intact. Topology proves that nature must cheat. In order to deal with the twisting and supercoiling, the ladder sides must some-how pass through each other during the separation process in order to make the snarled ball of DNA less knotted. And indeed nature does cheat: During DNA replication, the ladder sides are severed and then reattach themselves after passing through the other side— a highly illegal move in a topological world. The cutting and gluing is done not with knives and tape but by incredible enzymes that untangle the DNA and let each half live its own separate life.

Playing with abstract ideas within a fictitious amorphous universe thus leads us to amazing insights into our more rigid real world. Topological issues are clearly at work within the most foundational aspects of our lives. After our dip in the gene pool, we dry off and leave this microscopic realm and close the chapter with a variety of physical experiments that pique our curiosity and further challenge our intuition.

A TWIST OF FATE

We often hear that there are two sides to every issue. Here, though, we discover an elegantly twisted world where we expect two sides—

but find that there is only one. This world, which in fact is a twisted strip closed up to make a loop, is known as a *Möbius band*. It enjoys a celebrated and distinguished history; its attractive curves grace museums and even countless recyclable plastic containers. Here, in our final taste of topology, we will challenge our intuition about twisted worlds through several simple experiments with paper and scissors. We will uncover the nuance and subtlety of a single twist.

We begin by building a Möbius band—and we strongly urge you to participate by making one of your own. Take a strip of paper approximately 11 inches long and 2 inches wide, and hold the two narrow ends together to create a squat cylindrical loop (*Figure 9.31*).

Fig. 9.31

Notice that it has two sides (an inside surface and an outside surface) and two edges (a top edge and a bottom edge). This shape may remind you of the label that wraps around a tuna-fish can. All very ordinary and familiar.

But one theme of this book, as you've probably noticed, is that intrigue is just a twist away from the mundane. So let's make a twist—literally. We simply turn one end over, placing a twist—or, more accurately, half a twist—in our strip. We seal the ends together with tape and gaze upon our creation (*Figure 9.32*). The object that we now hold is a Möbius band. It is elegance, grace, beauty, charm, intrigue, and mystery all twisted together.

Fig. 9.32

A ONE-TRACK MIND

How can one paper loop contain all these qualities? Let's start our exploration by dipping a finger in india ink (or finding a felt-tip marker) and then slowly running it along an edge of the band so that the edge soaks up ink. When we return to our starting point we are surprised to see that we have inked up all the edges—that is, we find no ink-free edge (*Figure 9.33*). It seems commonsensical that a band would have two edges—and our band certainly did when it still looked like a tuna-fish label—but the Möbius band defies reason and possesses only one.

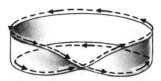

Fig. 9.33

Now consider the sides of the Möbius band. Using a pen (or ink on a finger) and starting on the outside, we trace a line around the band near the center (*Figure 9.34*). When we finally return to the place where we started, we notice another strange phenomenon: We have traversed "both" sides without lifting our pen (or, if we're using india ink, without lifting a finger). What we have actually established is that there is only one side!

Fig. 9.34

A somewhat more literary manner in which to confirm this fact is to take a strip of paper and on one side write "A MÖBIUS BAND HAS ONLY ONE SIDE; IN FACT, IN" so that it spreads across the entire length of the strip. Now flip the strip over by turning the bottom edge over the top edge, and write on this side "*COINCIDENCES, CHAOS,*

AND ALL THAT MATH JAZZ I DISCOVERED THAT" (and don't use a period). If we create a Möbius band by taping the right edge to the left edge with a half twist, then we have penned an inspirational true story that has neither a beginning nor an end but covers the entire Möbius band. This unity of edge and face is a surprising discovery that, as we will now see, leads to other intriguing properties.

MAKING THE CUT

The Möbius band displays beauty, and its beauty is not simply skin-deep—by cutting it open we find further hidden charms. With a pair of scissors, suppose we cut the strip in half lengthwise—that is, around the center of the band —until we return to our starting point (*Figure 9.35*). Ordinarily, when we cut an object in half, we produce two pieces. But as we are discovering here, the Möbius band is far from the ordinary.

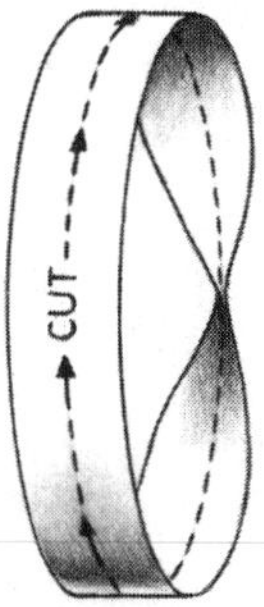

Fig. 9.35

The only way to fully appreciate the experiment is to try it. Make a Möbius band, and cut it lengthwise right in the middle. The result is a surprising single long strip that now has two twists. How can we deepen our understanding of the Möbius band so that this startling outcome makes sense? In other words, how can we become so intimate with the seductive Möbius band that its one-sidedness, one-edgedness, and one-piecedness-after-cutting all become obvious?

SOME ASSEMBLY REQUIRED

One potent mindset through which to look at the world is to notice how it is constructed. In the case of the Möbius band, its construction can be described by an *identification diagram*, which represents an object in its preassembled state. An identification diagram consists of a rectangular strip of paper with instructions explaining which edges are to be glued together to create the object. The assembly instructions consist of arrows drawn on pairs of the edges. Those edges are to be glued together in such a manner that the arrows point in the same direction.

Before we face the Möbius band, let's look at how we would create the identification diagram for a simpler object (*Figure 9.36*). If we draw arrows on one edge and arrows pointing in the same direction on the opposite edge, and then if we glue those edges to each other so that the arrows align, we will have constructed a squat paper cylinder like that tuna-can label. How do we produce the analogous identification diagram for the Möbius band?

Fig. 9.36

The answer is to simply draw the arrows on the edges pointing in opposite directions (*Figure 9.37*). If we bring those edges together with a half twist so that the arrows all point in the same direction and glue the edges together, then we will have constructed a Möbius band. But let's not actually take that last step just now. Instead, let's simply study the loose, unglued strip of paper, our unassembled

Fig. 9.37

Möbius band. Given the gluing instructions implied by the arrows, we see that as we move off the right edge at a point, we instantly pop out at the associated point on the left edge (*Figure 9.38*). In some

Fig. 9.38

sense we're traveling eastward on a twisted PacMan screen. So if a PacMöbiusMan were to run off its unassembled Möbius band on the upper right edge, he would instantly appear on the lower left edge (*Figure 9.39*).

Fig. 9.39

The identification diagram enables us to see the features of the Möbius band more easily. For example, if we travel eastward along the top edge, eventually we get to the upper right corner. That corner is attached to the bottom left corner when the edges of the Möbius band are glued together. So as we proceed, we see that the bottom edge of the unassembled band is really just a continuation of the top edge. When we get to the bottom right corner, it will be attached to the upper left corner, and from there we proceed back to the starting point after having traversed both the top and bottom edges of the strip of paper (*Figure 9.40*). Thus we have proven geometrically that there is only one edge to the Möbius band, just as we noticed before with our inky finger.

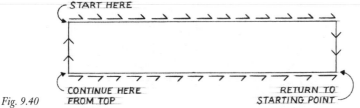

Fig. 9.40

A CUTTING PERSONALITY

The identification diagram also shows us why we are left with only one piece after cutting the Möbius band along the center (*Figure 9.41*). To see this phenomenon, we simply ask an ant to walk east-

Fig. 9.41

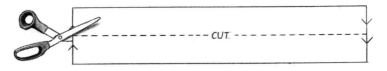

ward along the unassembled top half toward the right until she comes to the right edge (*Figure 9.42*). The very next step puts our ant on the bottom portion of the leftmost edge of the unassembled band. Our ant continues until she arrives at the lower rightmost edge and then steps out onto the top left edge and returns to her starting position. Since our ant traveled along both the top and bottom halves of our cut band, she verified that a Möbius band cut in half remains in one piece.

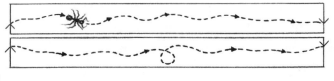

Fig. 9.42

To tickle your fancy and further challenge your intuition about the Möbius band, we pose the following question: What would happen if we cut the band lengthwise, but this time, instead of cutting right along the center, we cut it while staying a fixed distance of, say, $\frac{1}{4}''$ from the edge (*Figure 9.43*)? This experiment is truly terrific. When you try it, don't give in to the strong temptation to deviate from hugging that edge—force yourself to stay the course, and never cross the center! Until you've attempted it for yourself, avoid glancing at the footnote, where the outcome of this cutting feat is revealed. Once we've completed the experiment, if we consider how the cut appears on the identification diagram for the Möbius band, then we can see why that surprising outcome is less surprising (*Figure 9.44*).

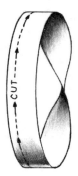

Fig. 9.43 Keep the scissors to the left edge.

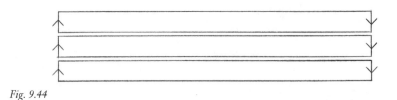

Fig. 9.44

The surprising result is two linked bands, one twice as long as the other.

Once we become conscious of the Möbius band, we suddenly see it everywhere. In fact, we find it on the bottom of nearly every plastic or glass bottle or cardboard box that can be recycled (*Figure 9.45*). So not only is the Möbius band intrinsically fascinating, it is also environmentally sound.

Fig. 9.45

ROLLING OUT THE DOUGHNUTS

Another important life lesson that mathematical thinking offers us in our everyday lives is a process for generating new ideas and becoming more creative. That procedure simply entails taking a notion and considering variations on the theme. In this context, for example, we created an identification diagram for the Möbius band by indicating how the two short edges of a rectangle were to be attached to each other. With the Möbius band as our inspiration, we can now consider different edge identifications for a sheet of paper and thereby discover and create new and exotic surfaces.

Suppose, for example, we identify the two sets of opposite sides of a rectangle (two horizontal sides and two vertical sides) and put

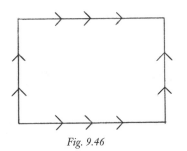

Fig. 9.46

arrows facing in the same direction for each of the pairs (*Figure 9.46*). If we curve the rectangular sheet so as to align the top and bottom arrows and then glue the edges together, then we will have created a tube. Notice that this tube has two circular edges marked with arrows. If we now bend the tube so that the arrows on the circular edges match, then we will have made the delicious surface of a doughnut (*Figure 9.47*).

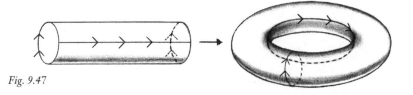

Fig. 9.47

A BOTTLE WITH A TWIST

Given this taste of success, why not try gluing the edges of a rectangle together in a different way? This time we will construct a still more intriguing object, one that combines the twisted turbulence of the Möbius band with the tubular tranquility of the doughnut. To begin, let's simply modify the identification diagram for the doughnut by switching the direction of the arrows on one of the sides (*Figure 9.48*), as we did to make the Möbius band instead of the tuna-can label.

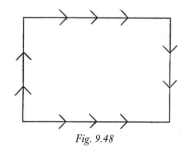

Fig. 9.48

Now we glue. Gluing the top edge to the bottom edge still produces a tube. However, as we bring the two circular ends of the tube together, we are in for a rude awakening: The arrows at those two ends do not align (*Figure 9.49*). One set of arrows is going around its circular end in one direction, while the other set is traveling around *its* end in the opposite direction. So how can we match up the arrows and finish our gluing?

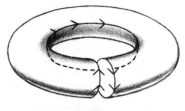

Fig. 9.49

ARROWS DON'T ALIGN

As we sit in frustration holding up one end of the tub in our right hand and one end in our left, we suddenly notice that now, when the ends of the tube are side by side pointing up at the ceiling, all the arrows are going in the same direction (*Figure 9.50*). But when we bring our hands together to make the ends meet, one set of arrows seems to flip—the arrows no longer point in the same direction by the time the ends touch. We try it again. As long as the ends of the

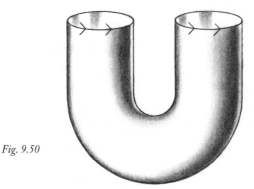

Fig. 9.50

tube point up at the ceiling, the arrows point in the same direction, but when the ends face each other, one set of arrows is wrong. How can we make the ends meet without inadvertently flipping one set of arrows in the process?

The answer is to cheat. Let's cut a hole in the sidewall of the left-hand part of the tube as we are holding it upright so that the portion we are holding in our right hand can be inserted (*Figure 9.51*). We push up until the top of the inserted portion meets the top of the other part of the tube. Now the two sets of arrows are pointing in the same direction as the edges meet, and so we can glue them together as specified. On the bright side, we have just constructed what is known as a *Klein bottle*; on the dark side, we confess that it's actually a Klein bottle with a hole, since we had to break into the tube. But back to the bright side: There is no obstacle to our imagining a Klein bottle without a hole and thinking about its properties.

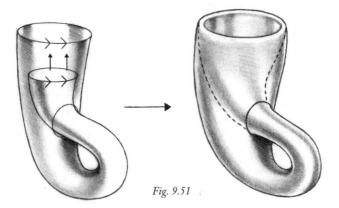

Fig. 9.51

Certainly a sealed bottle of champagne has an inside and an outside, but it turns out that the Klein bottle is a sealed bottle that has neither an inside nor an outside. Klein bottles have two small defects—one, any champagne put "into" such a bottle would immediately lose all its bubbles, and two, a Klein bottle is an abstract object that we can't completely construct in our constrained, real world. Those minor deficiencies aside, Klein bottles are the cat's pajamas and the bee's knees of alluring mathematical shapes.

TAKING SIDES ON THE KLEIN BOTTLE ISSUE AND UNFOLDING OUR UNIVERSE

The Klein bottle is quite elegant. Let's explore its ins and outs by placing our ant on the outside of the bottle and inviting her to take a stroll *(Figure 9.52)*. Having decided to tour the region that slightly resembles Niagara Falls (A), she travels downward. Slowly she feels a claustrophobic sensation as she travels through a tube-like hallway (B); then she breathes a sigh of relief as the narrow hallway widens out (C). When she reaches the ceiling (D) we suddenly realize that

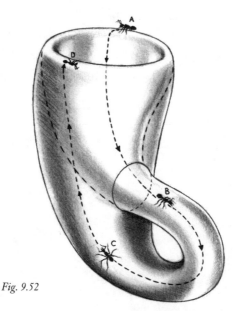

Fig. 9.52

she is now on the opposite side of where she began. Thus we are faced with the unusual circumstance of a sealed bottle that possesses only one side. The one-sidedness of the Klein bottle should not really surprise us, since we know from the identification diagram that it contains the one-sided Möbius band.

The Klein bottle, the Möbius band, and the surface of a doughnut are extraordinarily beautiful mathematical objects. These topological creations seem rather fanciful and abstract (aside from that doughnut!), but they are far from useless. The method by which we construct these mental playthings can be applied to the cosmic task of envisioning the possible global structure of the universe. The strategy is to use higher-dimensional analogues of the identification diagrams that we used for creating Klein bottles and doughnuts to create models of the universe. Here's how.

Instead of starting with a square, as we did in the construction of the Klein bottle or doughnut, let's start now with a cube. A cube has six faces. Suppose we pair opposite faces with each other in a manner analogous to how we paired opposite edges of a rectangle to describe the construction of a doughnut. We can imagine gluing those three pairs of opposite faces of the cube together. If we had a very elastic rubber cube, we could stretch it out and glue the right and left walls together, creating a squarish rubber doughnut (*Figure 9.53*). Unfortunately, we cannot physically perform the remaining two prescribed gluings. Instead, let's simply consider the object we would create if we could do all three. If we walked through a face of the cube, we would emerge through the opposite face—we'd be living in a three-dimensional PacMan

Fig. 9.53   Generalizing the surface of a doughnut by gluing opposite faces of a cube.

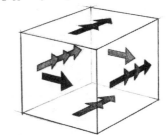

One pair of the faces glued together—gluing the other faces is a great challenge.

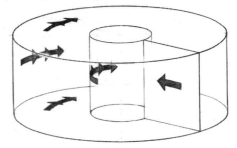

world. If we floated up through the ceiling, we would arise out of the floor (*Figure 9.54*). If we looked through one wall, we would see the

Fig. 9.54   An initially disturbing view until we realize that the ceiling is also the floor.

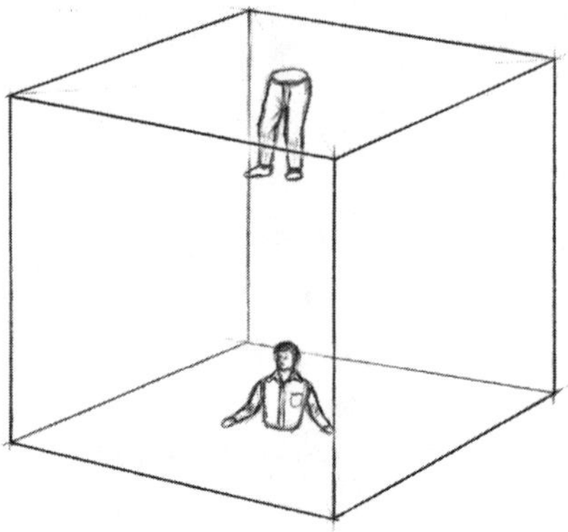

back of our own head. In fact, we would see infinitely many copies of ourselves receding into the distance (*Figure 9.55*).

Astronomers wonder whether this type of phenomenon actually

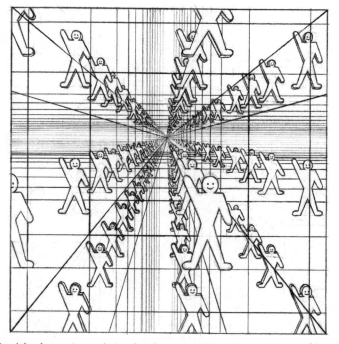

Fig. 9.55   A doughnut universe: A gingerbread person would look forward and see the back of its own head. This spaces appears endless and infinite, but it's really just a cube. Is our own universe truly endless?

happens in the real night sky. Might we be seeing the same star or galaxy in two opposite directions? The cube with identified faces is just one possible model for our universe. By specifying identifications with twists, we could create models of three-dimensional worlds that have properties like those of the Klein bottle. Perhaps in reality we ourselves are ants unwittingly strolling through a Klein-bottle universe.

PATCHING THE PUNCTURE BY PACKING OUR BAGS

While our discovery of the Klein bottle warms our insides (or is it our outsides?), we are plagued by a problem. In order to physically create that one-sided surface, we had to puncture the tube. Can we avoid having our surface pierce itself? The answer, sadly, is no—in reality we cannot. However, what if we were no longer constrained by three-dimensional geometric reality? Then the answer changes to yes—we can avoid puncturing the tube. In the next chapter we explore a world that transcends our own. We will journey into the fourth dimension. That spacious playground within our imagination will offer us an endless array of fascinating surprises, will provide a new vantage point for our everyday world, and will even permit an unpunctured, perfect Klein bottle. We leave our tour of the amorphous world of rubber-sheet geometry with a sense that the stretchable realm of topology frees our intellect to see ourselves and our world from different points of view and with more flexibility.

PART IV

TRANSCENDING REALITY

The Fourth Dimension and Infinity

Our last three chapters take us beyond reality. Mathematics is not constrained by mundane reality. It can build castles in the air and concepts in the mind whose beauty, magnificence, and intrigue are as boundless as the ideas themselves. Here we will explore two transcendent mathematical domains that have captured the imagination of curious minds throughout the ages: the fourth dimension and infinity.

The fourth dimension is an idea, a creation of the mind that is built on the solid foundation of real experience. The fourth dimension arises when we construct a new universe by answering the question "What if?" What if it were possible to move in a direction that is beyond our daily experience? Can we stretch our intuition to include a concept that is a creation of the imagination alone? We can. We can create a coherent concept in which we can explore, explain, and examine questions and possibilities that move us beyond the bounds of geometric reality.

Infinity arises from the question "What's next?" What number is beyond all the ones we can count? Can we make reasonable sense of the infinite? The answer is yes. By exploiting a simple, childlike concept, we will be able to stroll comfortably among the giants of infinite size. We will see infinity by thinking about our finite experience and then transporting ourselves beyond those bounds to embrace sizes that eclipse every actual number.

We will end our journey by breaking beyond the final frontier of infinity itself, discovering that our whole concept of infinity must grow to include the amazing insight that infinity itself is not self-constrained. There must be infinities beyond infinity. We will find them and enjoy the grand panoramic view of an idea truly without bound.

The vehicle that takes us beyond our own world into the worlds of coherent imagination is the powerful tool of focusing on simple ideas—notions so basic that they often go unnoticed and unexploited. By looking at simple, everyday experience, we find those sparkling clear crystals of insight that allow our minds to create worlds that no human will ever see but that we can imagine and explore. Understanding simple things deeply is the key that creates new dimensions and new grandeur for us to seek and enjoy.

———

CHAPTER 10

THE UNIVERSE NEXT DOOR

The Magic of the Fourth Dimension

> *. . . listen:there's a hell*
> *of a good universe next door;let's go*
>
> —E. E. Cummings

Obviously . . . Suspense builds as the spotlight shines on a safe submerged in an enormous glass tank of water. Crammed inside the soon-airless safe lies the charismatic illusionist David Copperfield. As time ticks away and water seeps into the safe, we know that this impossible escape act threatens to do in Mr. Copperfield just as it threatened to do in (and may have indirectly done in) Mr. Houdini. Finally, the safe is removed from the tank and unlocked. As its door opens, water floods out onto the stage. The audience gasps: The vault is empty. Wild applause erupts as David Copperfield suddenly appears from the wings in a perfectly pressed tuxedo. Plainly, this illusion alone was worth the price of admission.

Surprise . . . If we could use the fourth dimension—a new direction that would offer an extra degree of spatial freedom—then any one of us could easily escape from the vault and reappear in a rented tux.

Mr. Copperfield's illusion would be, in short, trivial. In the fourth dimension, all magicians would be out of work, and David Copperfield would become the charismatic fellow who flips our burgers. The intriguing world of the fourth dimension opens many doors for us to explore but closes many doors for aspiring illusionists.

DO YOU BELIEVE IN MAGIC? HOW ABOUT THE FOURTH DIMENSION?

Everyone has heard of the fourth dimension, but what exactly is it? Is it science fiction? Is it time? Is it where my keys have suddenly disappeared to? As we'll see, the world of four dimensions is magical indeed: Rabbits could materialize inside sealed boxes; we could escape from locked handcuffs with a wave of our hand; and we could be maximally ambidextrous—we could actually reverse our right and left sides. Exploring this magical universe not only is intriguing in its own right but also offers us insights into our real 3-D world.

Of course, it's possible that our actual universe is more magical and less mundane than we believe. It's possible that we live with an extra dimension but just can't find it. Maybe, like our keys, it's hidden under the cushions—perhaps down and to the left. But regardless of the reality of extra hidden dimensions, merely exploring the *idea* of the fourth dimension helps us break through the confines of our narrow personal experience and conceive of wonderful unseen worlds just barely out of view.

So what is the fourth dimension? Actually, this question raises a more fundamental issue: What is a dimension? In order to answer that, we begin by looking at the world around us.

DEGREES OF FREEDOM—VENTURING BEYOND OUR OWN SPACE

Physically, our freedom of movement seems constrained by the space we see. We can move back and forth, side to side, and up and down (*Figure 10.1*)—three different directions that in combination

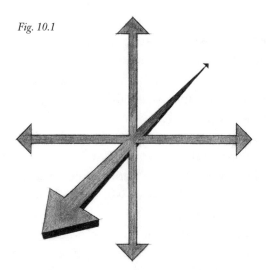

Fig. 10.1

can lead us anywhere we wish to journey. Thus around us we perceive a *three-dimensional* world. Dimension, in a vague sense, represents degrees of physical freedom. For better or worse, three degrees of freedom seem to be about all we have.

But let's be more precise and view dimension as having to do with the number of directions required to pinpoint any location in terms of a fixed reference point. For example, suppose we wish to specify the location of a slumbering mosquito perched on a hanging light bulb in a room, using a certain corner as our reference point. Then from that corner we can reach that blood-sucking houseguest by traveling 3 feet east, 4 feet north, and 7 feet up—that is, we can pinpoint our pest's precise location by giving its coordinates in three-dimensional space (*Figure 10.2*).

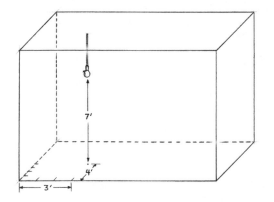

Fig. 10.2

As always, the best way to make sense out of a new idea is to consider the simplest possible illustrations and then move to the complex. Our three-dimensional world, while certainly familiar, is far from simple. So for the moment, let's retreat from our three-dimensional world and instead explore even simpler worlds—worlds with fewer dimensions to play with.

We could consider a two-dimensional flat, tabletop world, or even a one-dimensional line world. But let's start even lower. The world with the fewest possible dimensions is a zero-dimensional space. In this utterly freedom-free world, no directions are required to pinpoint a location. But if we don't need any specifications to find any particular place, then there must be no choices of locations; in other words, there is only one location in the entire space. So a zero-dimensional world is simply a single point (*Figure 10.3*). There are no degrees of freedom—in fact, we couldn't move at all. If we lived in a zero-dimensional world, we'd be home by now and we'd never miss a party. Even with those advantages, a zero-dimensional universe is, well, somewhat confining, so let's add a dimension and give ourselves a bit of room.

•

Fig. 10.3

A one-dimensional world comes next. We could see it as a line or, perhaps easier to picture, as one infinitely long street: The whole one-dimensional universe consists of one street that continues on forever. If we view this street as a number line (*Figure 10.4*), then locating anyone in this universe requires only one piece of information—namely, the address number (*Figure 10.5*).

Fig. 10.4

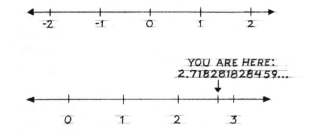

Fig. 10.5

YOU ARE HERE:
2.718281828459...

HOME ON THE PLANE AND BEYOND

Adding another dimension, we arrive at a two-dimensional universe, which can be pictured as a *plane* or the surface of a table. In the two-dimensional universe, there are two degrees of freedom: north-south and east-west. In other words, to pinpoint anyone in the plane we need two pieces of information. From a central starting point known as the origin, we need to specify how far north or south to go and then how far east or west. The north-south information takes us to the correct street; the east-west number takes us along the street to the house (*Figure 10.6*).

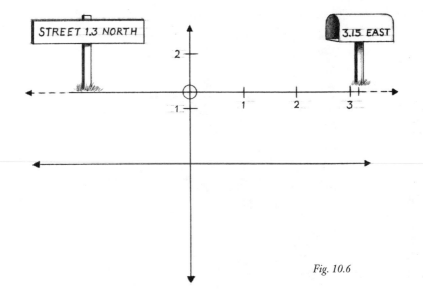

Fig. 10.6

If we now move up to our everyday world, we realize that the space we see all about us is three-dimensional, since we need a bare minimum of three pieces of information to precisely pinpoint a person. If new acquaintances tell us they live at 125 West 57th Street, we would still not know exactly where they are—they could be in the penthouse apartment or in the basement or anywhere in between (*Figure 10.7*). We need one extra piece of information—in this case, their floor. The north-south, east-west, and up-down directions are all needed in our three-dimensional world.

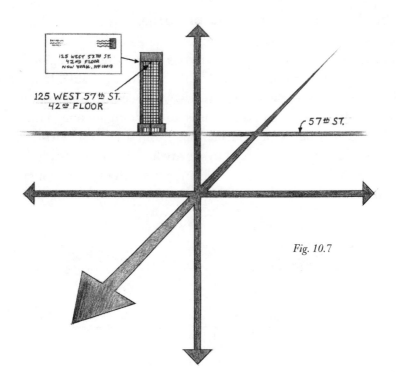

125 WEST 57th ST.
42nd FLOOR

57th ST.

Fig. 10.7

So what is a four-dimensional world? It's easy to say—it's a world that requires exactly four pieces of information to precisely pinpoint any point in that world. Okay, but correct though that statement is, it doesn't provide any real insight into such a mysterious world. So let's build up our intuition by building up worlds and then see if we can work our way to the fourth dimension.

INKING OUR WAY TO THE FOURTH DIMENSION

We can think of a line as a continuum of points strung together. That is, if we were to ink up a point and drag it in one direction for a while, then we would be sweeping out a line (*Figure 10.8*). So this

Fig. 10.8   Dragging an inked-up point sweeps out a line.

one-dimensional space is simply a zero-dimensional space that has been inked up and dragged. Or, without the messy metaphor, one-dimensional space can be viewed as a densely packed collection of zero-dimensional spaces.

If we now ink up the line and drag it in a new direction, then the ink will sweep out a plane (*Figure 10.9*). Equivalently, we can think of the plane as a tightly stacked collection of lines. Moving to the

Fig. 10.9

next level, we see that by analogy we can view our own three-dimensional space as an inked-up and dragged copy of one plane (*Figure 10.10*) or as a densely stacked collection of parallel planes.

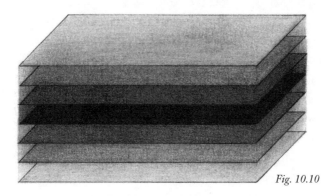

Fig. 10.10

We could casually consider three-dimensional space as an enormous ream of paper (*Figure 10.11*). Each sheet resembles a plane and has no thickness, but when they are stacked one on top of the other, they sweep out space—a huge rectangular block of paper that is a joy to

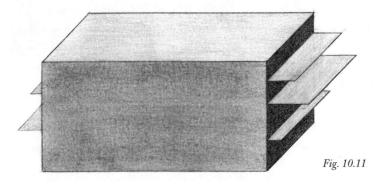

Fig. 10.11

behold (especially if you have ever had to reload the paper in a copier machine).

This visible pattern leads us to invisible worlds. Now we can ink our way up to the fourth dimension. What do we do? We simply take all of three-dimensional space, ink it up, and drag it in a totally new direction— one that we cannot see, since the only directions we perceive around us are contained in our three-dimensional world. Alternatively, but perhaps no better, we can view four-dimensional space as a bunch of three-dimensional spaces stacked up (*Figure 10.12*). This image is not easy to imagine either—since three-dimensional space is all around us, how could we stack it and not just get more of the usual space? The problem is that we have to stack or drag (if you're drawn toward ink) in a completely new and different direction from the current ones in our space—and what would that mean?

So the four-dimensional space that we just constructed—or tried to construct—is almost meaningless at this point. It is an abstraction that we cannot yet truly grasp. How can we come to understand this alien four-dimensional world? Usually we adopt the strategy suggested in the saying "If you

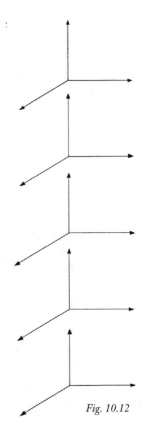

Fig. 10.12

want to know someone, walk a mile in his or her moccasins," but, unfortunately, we don't know anyone who has four-dimensional moccasins. Walking around in four-dimensional moccasins is such a foreign idea that we don't know where to start. So what do we do? We give up for now on the fourth dimension and retreat to the second.

A two-dimensional world is still a science-fiction fantasy world to us, but it has one major advantage over the fourth dimension: It is a *simpler*—rather than a more complex—foreign world. We'll soon discover that strolling around in the commonplace tabletop world of two dimensions introduces us to creatures whose bodies and minds enable us to better understand our 3-D selves and even the fourth dimension. This journey to both simpler and more complicated science-fiction worlds will bring us to the insight that our own familiar three-dimensional universe is itself a science-fiction fantasy from the vantage point of other dimensions.

HERE'S LOOKING AT YOU, (2-D) KID

Let's imagine a two-dimensional universe whose entirety is the plane surface of this paper. We imagine creatures populating this world, living out their lives unaware of the existence of any pages that came before or any pages that will follow this one thin page. This exercise of the imagination helps hone our abilities to grasp foreign domains. What are the consequences of the two-dimensionality of this paper-thin world? How would the inhabitants look? What could they see? What would they eat? Where would they bank? How would they look to us?

We'll begin by creating a two-dimensional creature—let's call him Slim—and giving him a home somewhere near the last word in this sentence. How would Slim look to us as we look at the page, his two-dimensional world, from our three-dimensional vantage point? Feel free to dabble at a doodle if you desire.

We might expect to see a face smiling up at us, maybe looking something like the image in Figure 10.13. We might expect that, but we'd be wrong. Why? Recall that Slim's entire world is this sheet of paper. If Slim did resemble that smiling face, then what would *he*

Fig. 10.13 *A natural (albeit incorrect) guess as to how a two-dimensional world might appear to us.*

see? Notice that all he would see is the inside area of his head; he could not see outside his head to the planar world around him, and of course the world above him (from where we are looking down) does not exist for him (*Figure 10.14*). That is, his eyes are located *inside* his body. Where are our eyes located? We find them along the interface between our internal body and the external world. Slim's eyes need to be located at his outer edge so that he can see out into his world.

What about Slim's mouth? Unless he plans to consume himself, good old Slim is in trouble—for food in his two-dimensional world is blocked from his mouth by his skin (*Figure 10.15*). In fact, his skin—the circle that defines his outer edge—forms a boundary that prevents any interaction between his mouth and his outside world. Our mouths are not located deep inside

Fig. 10.14 *If our eyes were totally surrounded by the boundary of our skin, we'd see only our inner organs—not the most desirable view for those of us who want to be outward-looking.*

our bodies, but in the more practical location along our outer surface. Just as Slim's one-dimensional skin forms a boundary that

Fig. 10.15   His skin would block two-dimensional food from reaching his mouth. (Who needs the Atkins Diet?)

keeps his internal organs in and his two-dimensional external world out, we now notice that our two-dimensional skin acts as an analogous boundary in our three-dimensional world.

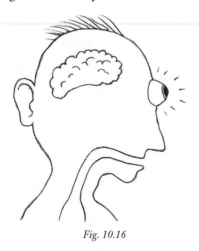

Fig. 10.16

Armed with these new insights, we now appreciate that our view of Slim's world is far different from *his* view of his world. For Slim to survive in his world, his body parts can't be where we first guessed. His mouth, his eyes, and an ear or two have to be placed along his boundary (*Figure 10.16*), because otherwise he cannot use them to gather food or information. Notice that we have a huge advantage over Slim: From where we are, we can see all the objects on the page, their insides as well as their outsides. Nothing in the two-dimensional plane, in fact, can be hidden from the sight of our three-dimensional eyes.

By analogy, we now realize that a four-dimensional being—let's call her Dee—would have a similarly richer view of our world than we ourselves have. That is, nothing in our three-dimensional slice of Dee's four-dimensional world could be hidden from her sight. She could reach down and touch our inner organs without ever piercing our skin. Just think of the medical possibilities if we could perform surgery with that extra degree of freedom! To illustrate this unusual perspective, we now turn back to magic and further develop insight into the fourth dimension by analogy (and perhaps some sleight of hand).

USING AN EXTRA DEGREE OF FREEDOM TO MAKE A RABBIT DISAPPEAR

We see the empty interior of an open box. It is then closed and sealed. Abracadabra—the box is opened to reveal a disoriented pink rabbit (*Figure 10.17*). How can this amazing illusion be performed using the fourth dimension?

Fig. 10.17

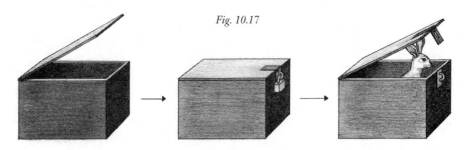

The mantra "Understand simple things deeply" is the key to unlocking mysteries. Whenever we're confounded by a question concerning the fourth dimension, we first consider analogous lower-dimensional versions. Retreating to lower, simpler dimensions can teach us how to understand the abstract fourth dimension. So instead of pulling a rabbit out of thin air using the fourth dimension, we retreat to the analogous disappearing act within the tabletop world of two dimensions where the third dimension does the trick.

A sealed two-dimensional box is simply a square (*Figure 10.18*). A square divides a plane into two regions, the inside of the square and its exterior, just as a box separates three-dimensional space into

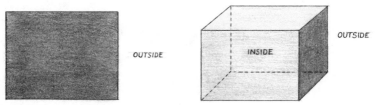

Figs. 10.18 and 10.19

two spatial regions, the inside and the outside (*Figure 10.19*). Now suppose two-dimensional creatures living in the plane keep their eyes on that sealed box and never look away (*Figure 10.20*). Is it possible for us, as three-dimensional beings, to use the third dimension to place a two-dimensional rabbit into that sealed box without our tabletop audience seeing it happen?

Fig. 10.20

Certainly it's possible—we can "airlift" the rabbit in. While our audience can move only along the surface of the table, we ourselves have the extra degree of freedom to move above the tabletop universe and have a panoramic, aerial perspective of that two-dimensional world (*Figure 10.21a and b*). Thus we can see the box

Fig. 10.21a

Fig. 10.21b

and both the inside and outside regions all at once—an impossibility for anyone living *in* the plane of the table. To us, the box is open; it exists as a square on one table-slice of our more robust three-dimensional world, and the area inside the square is just as accessible to us as the area outside the square. We could simply pick up a two-dimensional rabbit, bring it to a spot right above the box, and then place it down onto the tabletop inside the box. Would the audience see anything? Nothing but the sealed box, since our actions took place completely outside their flat world. Notice that the two-dimensional audience could not even point to the direction from which we lowered the rabbit. They will be amazed when they open the box and see a furry flat rabbit dazed and confused inside (*Figure 10.21c*).

Fig. 10.21c

Now we return to the original trick of using the fourth dimension to place a rabbit in a sealed box. We argue by analogy. Since four-dimensional space is composed of stacked layers of parallel three-dimensional spaces, then from that extra, fourth degree of freedom the interior of that sealed box is actually exposed, just as the inside of the square was visible from our three-dimensional view of the entire tabletop universe. Thus a four-dimensional accomplice could "airlift" the rabbit into that sealed vault from the direction of that new degree of freedom. The audience in the three-dimensional slice of four-dimensional space wouldn't see any funny business, but when they opened what appeared to them to be a completely sealed box, they'd discover, to their shock, a very real rabbit.

An extra degree of freedom allows anyone viewing a lower-dimensional universe to see things that cannot be seen by the inhabitants. The panoramic view makes it possible to see everything simultaneously—even things that, to the inhabitants of the lower-dimensional universe, appear completely sealed. It is counterintuitive to us, inhabitants of a 3-D world, that there might be a perspective from which it is possible to see the inside of a completely sealed vault and even to place an object inside it without our noticing. However, it is just as counterintuitive for those plane folk on the tabletop staring at the sealed square. To them, that square is completely sealed, while we, looking down at their tabletop world from above, see and can explore the inside of that square without ever touching its boundary. That one-dimensional boundary separates the two-dimensional tabletop world, but not our three-dimensional world, into the area inside the square and the area outside the square. Similarly, but less visibly to us, a sealed vault separates our three-dimensional world, but not a four-dimensional world, into the space inside the vault and the space outside the vault. In four dimensions, a sealed three-dimensional box resides in just a slice of space.

UNKNOTTING SHOELACES WITHOUT TEARS

Suppose we have a sealed loop of rope with a knot in it (*Figure 10.22*). It is impossible to get the knot out just by moving the rope around. The only way to unknot a real knot in a loop is to cut the rope, unknot the knot, and reseal the ends. If we could use the fourth dimension, however, then we could unknot any knotted mess without ever resorting to a pair of scissors.

Fig. 10.22

How? Again, let's retreat to a two-dimensional version of this question.

Actually, there cannot be knots in a tabletop universe: All objects in that universe lie flat on the plane, while knots require a third dimension—strands of rope have to cross *over* one another (*Figure 10.23*). But we can capture the spirit of the unknotting solution by

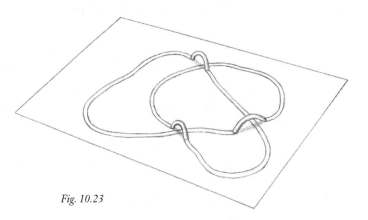

Fig. 10.23

considering a poisonous tabletop tsetse fly (*Figure 10.24*). In order to contain this monstrous mosquito, two-dimensional concerned citizens risk their lives to lasso that pest with some two-dimensional

Fig. 10.24

rope. They surround the tsetse fly with the rope, thus creating a loop in which the tsetse is sealed off from the rest of the plane. (This part of our story—and all future developments—can be viewed in comic-strip form in *Figure 10.25*.)

All is well until a three-dimensional terrorist decides to threaten the lives of all the citizens of the tabletop universe by setting the tsetse free. She knows, though, that the tsetse presents dangers even to three-dimensional terrorists, so she must accomplish her scheme without actually touching the fly. She hits upon the idea of picking up a piece of the rope that surrounds the fly.

What do the plane's people see? From their vantage point, a piece of rope suddenly disappears. The rope has not been cut, but to

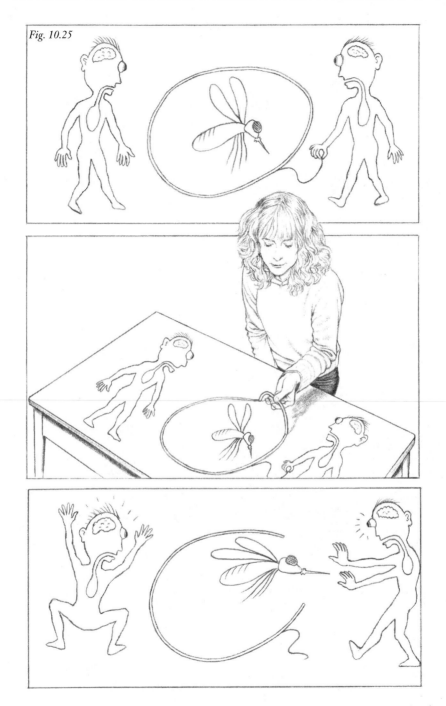

Fig. 10.25

. . . continued

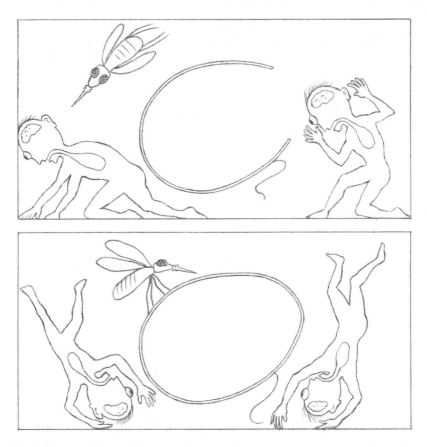

the citizens of the tabletop universe the rope now seems to be missing a segment. Of course, the rope looks that way to the tsetse fly also, and she is able to exit through the gap and terrorize the people of the plane again.

At this point the evildoer can now ease that bit of rope back down onto the surface of the table. To the inhabitants it seems that the cut rope has now mysteriously become whole again. In reality, as we know, the rope was never cut—it was just lifted into a new dimension, out of the tabletop inhabitants' sight. And so we come to the end of our sad tale—a frightening example of WMD (Weapons of Multiple Dimensions).

And now we're ready to tackle that loop of rope with a knot in it using the fourth dimension as a substitute for cutting. Let's use our evildoer's strategy in this higher-dimensional context. That is, we get a four-dimensional being to lift a piece of the rope into the fourth dimension. Then that segment of rope will be out of the three-dimensional space to which our sight is confined, and thus it will seem to us that this piece has been cut away (*Figure 10.26*). In

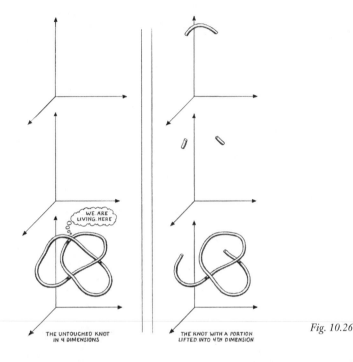

Fig. 10.26

reality, the rope remains intact, a segment of it simply having moved to parallel three-dimensional spaces that are invisible to us. In our three-dimensional world, however, the fact remains that the rope has a missing segment, and thus we can easily get the knot out (*Figure 10.27*). Now our four-dimensional friend lowers that segment

Fig. 10.27

back into our world; what we see is the illusion of rope ends fusing together (*Figure 10.28*). Suddenly the rope is a loop again—but happily, without that pesky knot. Using the fourth dimension, we unknotted without cutting!

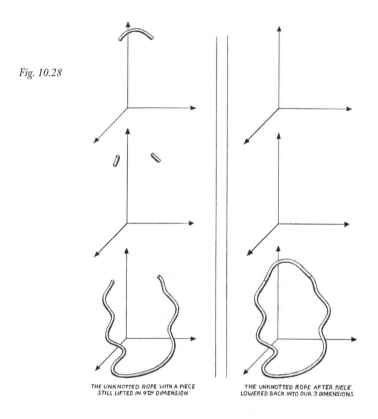

Fig. 10.28

THE UNKNOTTED ROPE WITH A PIECE
STILL LIFTED IN 4TH DIMENSION

THE UNKNOTTED ROPE AFTER PIECE
LOWERED BACK INTO OUR 3 DIMENSIONS

We can also apply this technique to patch up a problem from Chapter 9 involving the punctured Klein bottle. Recall that in order to assemble that beautiful one-sided, closed, bottlelike surface we were forced to cut a hole along its side so that it could pass through itself (*Figure 10.29*). But if we employ the fourth dimension, the Klein bottle's cosmetic surgery is no longer necessary. We merely lift a portion of the side of the tube into the fourth dimension (*Figure 10.30*). Thus, while there appears to be an open gap along the tube's side, in reality we understand that no cut was made—that missing-from-sight piece is hovering in the fourth dimension. We now can pass the tube through itself without making a hole, just as we unknotted the loop without cutting the rope.

A perfect, holeless Klein bottle cannot be made in three-dimensional space, because there are not enough degrees of freedom to connect the ends of the tube without a puncture. Thus we discover that the Klein bottle is an object whose natural home is the fourth dimension.

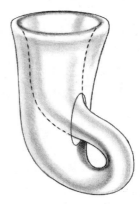

Fig. 10.29

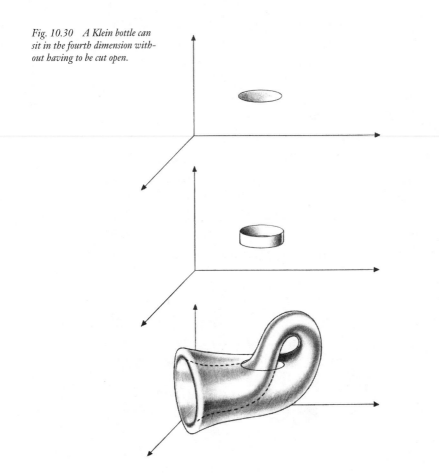

Fig. 10.30   A Klein bottle can sit in the fourth dimension without having to be cut open.

BUILDING CUBES BY INKING AND DRAGGING

Armed with the basic idea of building up from smaller dimensions to higher dimensions, we are able to tackle almost any dimension issue that we could possibly face. Just to illustrate our multidimensional abilities, we'll dabble in the geometry of the fourth dimension and discover how to construct a four-dimensional cube.

Where would we start? In grappling with the fourth dimension, we always start at ground zero. What is a zero-dimensional cube? Easy! In dimension zero, everything is simply a point, so a zero-dimensional cube is a point, a.k.a. a dot (*Figure 10.31*). How do we move from here to a one-dimensional cube? We ink up the dot and drag it one unit in a new direction, and we get a one-dimensional cube, a.k.a. a line segment (*Figure 10.32*). Now we're rolling. If we ink up that line segment and

•

Fig. 10.31

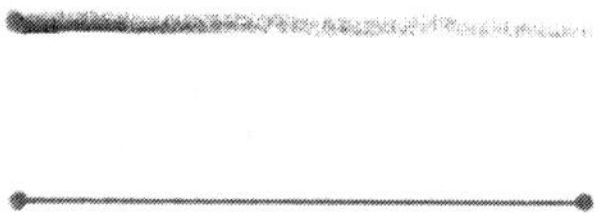

Fig. 10.32

drag it one unit in a perpendicular direction, we sweep out a two-dimensional cube, a.k.a. a square (*Figure 10.33*). Ink up the entire square (both its inside and its boundary) and drag it one unit perpen-

Fig. 10.33

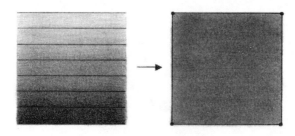

dicular to the others, and we produce a three-dimensional cube, a.k.a. a cube (*Figure 10.34*).

Fig. 10.34

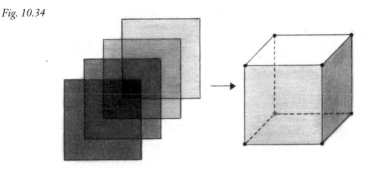

Now we see what to do next. We ink up the entire cube—even the points inside the cube, as if the cube were a sponge—and drag the entire cube one unit in a direction *perpendicular* to all the rest. Thus we have produced a four-dimensional cube—a.k.a., well, a four-dimensional cube (*Figure 10.35*). Of course, there is no need to

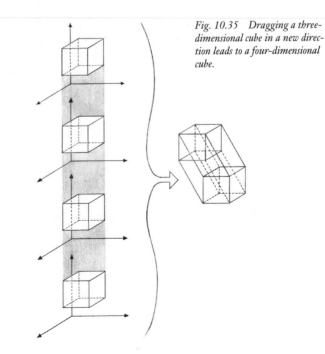

Fig. 10.35 Dragging a three-dimensional cube in a new direction leads to a four-dimensional cube.

stop here. As long as we have ink in which to dip and new directions in which to drag, we can generate cubes of higher and higher dimensions (*Figure 10.36*).

Fig. 10.36   A five-dimensional cube.

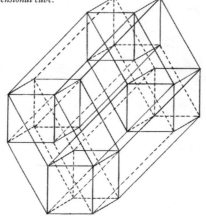

THE UNMAKING OF THE CUBE

Our drawings of three- and four-dimensional cubes suffer from distortion. For the three-dimensional cube, the faces are not perfect squares and the angles are not all right angles (*Figure 10.37*). What's

Fig. 10.37

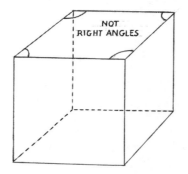

the problem? The answer is that we cannot generate an exact replica of a cube on a sheet of paper. The cube requires three dimensions while a sheet of paper can provide only two. Using perspective, we render some of the cube's angles and faces in a way that suggests the extra dimension. When we view the drawing with our three-dimensional eyes, we instinctively perceive that third dimension; in our minds we put the pieces together and see the cube as it was meant to be seen.

The four-dimensional cube is more challenging to visualize, since we don't have four-dimensional eyes to parse that complicated perspective shot. And note that a three-dimensional cube has only one dimension more than this page, so we have to compensate for only one dimension in that rendering, but a four-dimensional cube has two extra dimensions, and thus we require greater dimensional compression in our drawing. A more accurate rendering of a four-dimensional cube could be realized through a three-dimensional "picture." Although it's not possible to reproduce a three-dimensional model on a two-dimensional page, Figure 10.38 gives some sense of how the four dimensions would be rendered as three. In the model, the faces are not squared off and do not meet at right angles.

Fig. 10.38

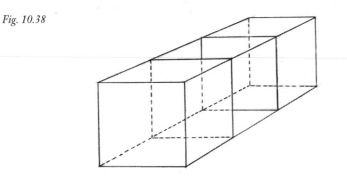

We can produce models of the boundaries of three- and four-dimensional cubes that possess perfectly square faces with all angles right. We capture them in a different way: Taking a lesson from the postmodern age, we deconstruct the cubes. If we unfold the boundary of a three-dimensional cube, then we produce six perfect squares all at right angles and joined together to form a cross (*Figure 10.39*). To assemble the cube, we need to associate pairs of edges of the squares and glue those pairs together (*Figure 10.40*). Similarly, if we "unfold" the boundary of a four-dimensional cube, then we discover

Fig. 10.39

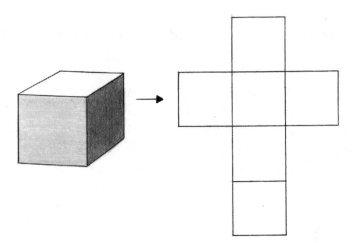

a collection of eight perfect cubes all at right angles and joined together to form a cross with two additional arms (*Figure 10.41*). The *faces* of the cubes can be glued together to assemble this collection of three-dimensional cubes into a four-dimensional cube (*Figure 10.42*).

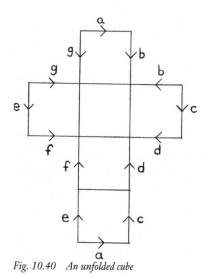

Fig. 10.40   An unfolded cube

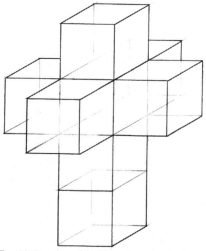

Fig. 10.41   An unfolded four-dimensional cube.

Fig. 10.42

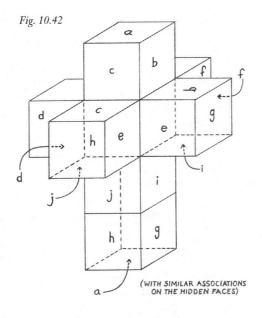

(WITH SIMILAR ASSOCIATIONS
ON THE HIDDEN FACES)

CAPTURING THE FOURTH DIMENSION ON CANVAS

The intriguing concept of representing a four-dimensional cube by showing its unfolded boundary was the inspiration for a 1954 painting by Salvador Dali, *The Crucifixion, Corpus Hypercubicus* (*Figure 10.43*), in which we see the fourth dimension taking on religious significance. The geometry and the sheer notion of the fourth dimension have inspired artists as well as scientists and mathematicians. In addition to Dali, the painters Marcel Duchamp and Max Weber were known for their explicit use of the fourth dimension in their works.

In Duchamp's *Nude Descending a Staircase #2* (1912), we see the entire span of motion captured at once, as if the nude herself were inked up and

Fig. 10.43

Fig. 10.44

dragged down the stairs (*Figure 10.44*). Thus Duchamp created a wonderful image that captures the totality of movement through one four-dimensional metaphor. His work leads us to wonder whether the fourth dimension can be viewed as time. We can view it as time if we wish. In fact, we can interpret that extra degree of freedom in a variety of ways, including sound or color. The difficulty with seeing time as a model for the fourth dimension is that we are unable to move backward and forward in it as easily as we can in the other three dimensions. Also, why should the fourth dimension be so radically different from the first three? By considering the fourth dimension as a spatial one we can develop insights into a potential geometry and reality beyond our own.

Max Weber's *Interior of the Fourth Dimension* (1913) offers the fourth dimension as an eerie, foreign world (*Figure 10.45*). Certainly we have seen for ourselves how foreign that extra degree of freedom is. However, now perhaps we can, through analogy, appreciate that the fourth dimension is a world that is just over our visible horizon. As our journey and the works of these artists will all attest, the fourth dimension is certainly a font of creativity, beauty, and wonder.

Fig. 10.45

LESSONS FROM THE FOURTH DIMENSION

The fourth dimension has a romantic, almost mystical allure. It seems to lie in the realm of science fiction, a world beyond the reach of our senses. In this chapter we tried not only to move toward the fourth dimension but to embrace and explore it. We embarked on our journey into the fourth dimension by quickly acknowledging that when we are faced with a difficult issue, it's often best not to tackle the challenge head on but instead to pick an easier target. Before advancing toward the fourth dimension, we retreated to our own three-dimensional world and then retreated farther still, considering a two-dimensional world and asking what our lives would be like if they were played out in a universe that consisted of the surface of a tabletop.

Developing ideas by systematic analogy is a fantastic method for creating new insights. These techniques can show us that our level of understanding, even of trivial matters, is not as great as we sometimes think. Looking at aspects of our familiar world from different vantage points can reveal new richness and surprise; and extrapolating from those new insights can fuel important, original ideas. Grappling with the fourth dimension teaches us to think way "outside the box"—taking us to an entire dimension beyond our everyday experience. Next, we bring our mathematical journey to a close by traveling outside of finite dimensions into the endless universe of infinity.

CHAPTER 11

MOVING BEYOND THE CONFINES OF OUR NUTSHELL

A Journey Into Infinity

I could be bounded in a nutshell,
and count myself king of infinite space.

—William Shakespeare

Obviously . . . We throw 10 Ping-Pong balls into a barrel; then we reach in, grab one, and toss it out. We don't have to be rocket scientists to see that 9 balls remain in the barrel. Now we throw 10 more balls in and take another one out. Even those who were taught the "new math" will agree that 18 balls remain in the barrel. Do it again and we are left with 27 balls; again and we have 36; again and we see 45; and so forth. We see the pattern. But what if we did it *forever*?

Surprise . . . The barrel would be empty. Welcome to the incredible and counterintuitive world of infinity.

———

For most of us, the closest we'll get to infinity in our everyday lives is thinking about Bill Gates's net worth. Even the word "infinity" evokes a grand sense of majestic vastness, mystery, and incomprehensibility. For most of history, people couldn't get beyond thinking of infinity as a fuzzy, awe-inspiring enigma, a godlike, all-encompassing everythingness, unknowable and unimaginable. People couldn't say many sensible things about it, but they did agree that infinity is big. In fact, they agreed that infinity is as big as it gets. They agreed—but they were wrong.

It turns out, as we'll discover in these final two chapters, that we can make perfectly good sense of counting to infinity—and beyond. In fact, we'll see that infinity is easier than pairing up socks out of the drier, for with infinity, there's no static cling. The journey to infinity, though not difficult, will lead us to counterintuitive revelations and insights.

UNKNOWABLE NUMBERS

The whole idea of infinity is that it is bigger than all numbers. It's certainly past 14; it's past 2,343; and it's even past 1,234,826. It's far bigger, actually, than

14,736,030,038,738,738,574,387,983,475,937,984,794,357,398,753.

In fact, on the road to infinity, that huge number is not even a speck of dandruff on a whale's scalp.

Before we face the daunting concept of infinity, let's first consider a seemingly easier question: Can we understand the familiar counting numbers, that is, 1, 2, 3, 4, . . . ? All's well until we face those three little dots after the 4, because the ellipsis tells us that we are to continue on forever, and forever is a long, long time.

As we saw in Chapter 4, we all understand 1, 2, and 3. We understand 2 because we see pairs of things all around us—socks, twins, dice. Likewise, 3 is intuitive to us because of familiar collections—tennis balls in a can, wheels on a tricycle, and the number of legs allowed in a three-legged race. As for 4, that's the number of bridge players, walls in a typical room, or hubcaps on a car parked in a safe neighborhood. But as the numbers flow, our intuitive sense of them ebbs.

As we saw before, the U.S. national debt is a number consisting of 13 digits. Specifically, on January 11, 2003, the national debt was clocked at $6,420,664,216,307. We have words to express that number: "Holy cow!" or more precisely, "Six trillion, four hundred and twenty billion, blah blah blah." It's difficult to have a real sense of the meaning of this enormous quantity.

Astronomers estimate that the total number of atoms in the universe could be written using about 80 digits—but do such spectacularly large numbers have any personal meaning to us? What about a number having a thousand digits? How about one with a million digits? Numbers with a million digits are essentially meaningless, yet most numbers are far larger still. Long before we reach infinity, we must face ideas beyond our grasp.

There is no way to make numbers that exceed the number of particles in our universe intuitive and familiar. Given these thoughts, how can we hope to understand infinity? As Woody Allen once lamented, "I'm astounded by people who want to 'know' the universe when it's hard enough to find your way around Chinatown."

INFINITY IS WITHIN OUR REACH

Happily, as we will now discover, the ascent to infinity is well within our mountaineering abilities. Indeed, the first surprise about infinity is that we *can* make sense of it at all. Our strategy is to back off from lofty concepts that are too big to handle and, instead, to think very carefully about easy, familiar things. Basically, when faced with a hard challenge, we turn and run the other way. In this case, instead of trying to understand infinity, let's try to understand 5. That's a notion we might be able to come to grips with, since it is the number of fingers on most hands.

The crucial observation comes into focus when we compare exemplars of the same number. In the case of 5, fingers on the left hand make up one example and fingers on the right form another. If we first have the tip of our left pinkie meet the tip of our right pinkie and then have our ring fingers touch each other, followed by our middle fingers, index fingers, and finally our thumbs, we see something simple but insightful (*Figure 11.1*). What we see is that there is a correspondence—a one-to-one pairing—between the fingers on

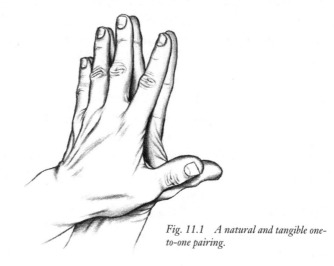

Fig. 11.1   *A natural and tangible one-to-one pairing.*

our left hand and the fingers on our right hand. Even if we couldn't count as high as 5, we would know for certain that however many fingers we have on our left hand, it equals the number of fingers on our right hand.

So now we have a new sense of 5. Five is the size of any collection of objects that can be paired up exactly with the fingers of a left hand. Thus the number of pennies required to equal a nickel is 5, since we can balance one penny on each of our fingers on our left hand with no pennies or pinkies left over. There is no resisting the natural and simple idea that if the elements of two collections can be paired up evenly—each item from one collection with an item from the other—then one collection contains the same number of things as the other. Within mathematics, such a one-to-one pairing between two collections is often referred to as a *bijection*. This shift in thinking from *counting* how many to *comparing* collections is the key to unlocking the infinite.

THE KEY TO INFINITY

Two collections whose contents can be put in a one-to-one pairing have the same size. Let's now look at the consequences of this innocuous idea and see where the "pairable means equal" concept leads us. With small collections such as fingers on the left hand compared with fingers on the right hand, or stars on a U.S. flag compared with states in the Union, pairable clearly means equal in number.

We're now ready to take a step into the unknown. Let's consider collections that contain *infinitely* many objects. Where would we find an infinite collection? Forgoing the finite confines of physical reality, we let our imaginations go wild.

WELCOME TO THE INFINITE INN. The next time you're traveling, forget about all those Hyatts, Hiltons, Omnis, Four Seasons, and Motel 6's. You've got to spend a night at the Infinite Inn, the hotel with the catchy slogans: "You can count on as many rooms as there are counting numbers"™ and "There will always be room at the Infinite Inn."™ The Infinite Inn truly deserves its name—its rooms are lined up and numbered 1, 2, 3, 4, . . . *forever*. There is one room for each counting number—now that's some hallway! Can we physically build it? Of course we cannot. Can we mentally build it? Of course we can. This suite scenario exists in our imagination—which we stretch to include an endless corridor of doors, each one leading into a modest but attractive hotel room (*Figure 11.2*).

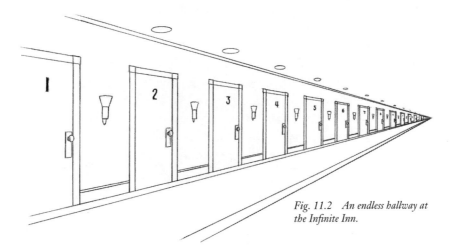

Fig. 11.2 An endless hallway at the Infinite Inn.

Certainly the first slogan is absolutely true—there is one room for each counting number. But does the Inn's truth in advertising end there, or is that second slogan true as well? Is it true that such an enormous hotel would never have need for a "No Vacancy" sign? Of course, in our physical real world containing about 6.4 billion people, the answer is yes, the slogan is true if there are only finitely many people, then the Infinite Inn cannot run out of rooms. There is enough space to allow every member of humanity—even those who are no longer living—to have his or her own private room, with infinitely many rooms left vacant. This incredible fact allows us to get a slightly better grasp of the Inn's size—all of humanity could check in and it would still appear essentially empty!

But if we are allowing ourselves to imagine an inn that is too vast to physically exist, then why stop there? Let's now stretch further still and imagine a scenario that is equally impossible—namely, a world of infinitely many people. In this new fictitious world, is the "There will always be room at the Infinite Inn"™ motto still true? Let's explore a few suite-filling scenarios and see if we can make our inn runneth over.

IS INFINITY PLUS ONE BIGGER THAN INFINITY? Let's suppose that the Saint Louis Cardinals expanded so that the team now includes infinitely many players, each sporting a numbered jersey: 1, 2, 3, 4, . . . *forever.* One day while on the road they decide to check into the Infinite Inn, which happens to be completely empty. Can each player get a private room? Certainly. The desk clerk puts Player 1 in Room 1, Player 2 in Room 2, and so on; that is, each player is given the room that has the same number as his jersey. Thus we've produced a one-to-one pairing between the players and the rooms (*Figure 11.3*). There are no players left on the bus, and

Fig. 11.3   A natural one-to-one pairing between the rooms and the players.

there are no empty rooms. Therefore, we see that in terms of their sizes, the collection of players equals the collection of rooms—no real surprise. Now, however, since all of the (infinitely many) rooms are filled, it appears that there is indeed a need for the desk clerk to flash the "No Vacancy" sign. He turns it on and then breathes a sigh of relief at having completed his work.

But now the owner of the Cardinals walks in, not pleased to be greeted by the "No Vacancy" sign. The clerk explains that the Inn is completely filled with players and that there are no empty rooms. "What about the very last room?" the owner shouts gruffly. The clerk, rolling his eyes, explains that there is no "last room," since there is an endless run of rooms—filled by the equally endless run of players. The manager of the Inn, hearing the commotion at the front desk, ventures out to find a frustrated desk clerk and an angry team owner. The manager calms them both down, reassuring them that she will be able to accommodate the owner while still giving each player a private room. Both the clerk and the owner are skeptical until the manager has explained her room-rearranging scheme thoroughly.

Here's how it works: She will ask each player to move out of his room and move over to the next room in numerical order. So Player 1 will move into Room 2, Player 2 will move into Room 3, Player 3 will move into Room 4, and so forth (*Figure 11.4*). It is clear that after the process is completed, each player will still have his own private room. Even Player 14890003862 has a room—which one? Room 14890003863, of course. Everyone is happy to oblige and shifts over at once. Now the manager notes that Room 1 is vacant and ready for the owner of the Cardinals to enjoy a peaceful slumber.

Thus we see a one-to-one pairing between the rooms, on the one hand, and the players *plus* the owner, on the other. So adding a

Fig. 11.4 *Everyone slides down by one room and still has a private room.*

new member to the infinite collection of players has not increased
the size of the collection—it can still be put in a one-to-one pairing
with the hotel rooms. Similarly, we now see that if another person
were to arrive—the coach, for example—we could just ask everyone
(including the owner) to shift down again; everyone would still have
a room to himself while freeing up Room 1 for the coach. In fact, if
100 more people were to arrive, we could shift everyone down 100
rooms. Thus, adding any finite number of elements to an infinite
collection does not increase the size of that collection.

IS HALF OF INFINITY SMALLER THAN INFINITY? Suppose now
that during a spirited baseball practice, half the team gets injured.
Oddly enough, those injured are precisely the players having odd-
numbered jerseys. So the players numbered 1, 3, 5, 7, 9, 11, 13, . . .
and so on *forever*, accompanied by the owner and the coach, are sent
home. But now half the hotel is vacant—in fact, every other room is
empty. While this development might be nice for those who value
peace and quiet, it is not ideal for fostering team unity and spirit.
Thus the players who remain at the Inn—that is, all those with even-
numbered jerseys—decide to move their rooms so that they can be
next to one another. Player 2 moves into Room 1; Player 4 moves
into Room 2; Player 6 moves into Room 3; Player 8 moves into
Room 4; and so on (*Figure 11.5*).

Fig. 11.5   *A one-to-one pairing between the rooms and the even-numbered players—all rooms*
are now filled.

Once they move, we notice something very peculiar. Are there
now any empty rooms? Well, Room 5 has Player 10 in it, Room 6
has Player 12, Room 22 is occupied by Player 44, and Room 23 is the
new home of Player 46, and so forth. Farther down the hall, we see
that Room 1031021 is occupied too—it houses Player 2062042.
Every room we look at—and every room we can imagine—has

someone in it. Unbeknownst to the even-numbered players, but now knownst to us, they have established a one-to-one pairing between the rooms and themselves. That is, after *half* of the elements of an infinite collection are removed, the size of the collection does not change!

The mathematical term *cardinality* is used to refer to size, particularly when speaking of infinite collections. "Two collections have the same cardinality" means that the elements of the two collections can be put in a one-to-one pairing. Mathematically speaking, we have thus shown that the collection of the even counting numbers has the same cardinality as the collection of all the counting numbers. Basically, half of infinity is no smaller than infinity.

IS INFINITY PLUS INFINITY GREATER THAN INFINITY? Let's really up the ante. Suppose that the entire Cardinals team is in the Inn again, having recovered from their injuries: Player 1 in Room 1, Player 2 in Room 2, and so forth. Instead of throwing in one or two extra people, let's throw in *infinitely* many more new people and see if we can overflow the hotel. Suppose now that another infinite team, the San Francisco Giants, arrives on the scene. The Giants, too, are numbered 1, 2, 3, 4, . . . *forever.* Can we use the shift-down method to give all these Giants their own rooms without tossing any Cardinals out on the street?

We can shift once to let the first Giant stay in Room 1, and shift the Cardinals again to let the second Giant check into Room 2. In fact, as we've already seen, we can shift any *finite* number of times to let in any *finite* number of Giants. But can we shift down an infinite amount to allow *all* the Giants to check in? Well, suppose we tried this infinite shifting. Then what is the room number of Cardinal Player 1? He cannot be in Room 1, or 2, or 3, or 4, . . . or, in fact, *any* room, since any room we can think of will be a *finite* number of rooms away from Room 1—and such a room is filled with a Giant team member. Thus Cardinal Player 1 is out on the bench and not a particularly happy ballplayer. At least he's with his teammates, though, since they've been thrown out as well.

So it appears that if we take a completely full Infinite Inn and add infinitely many more people, we overfill the place. Or do we? All we saw was that the "infinite shift" method does not work. But perhaps there is another pairing that will match rooms to members of both teams in a one-to-one manner. Let's try to modify the strategy used in the even-numbered player situation.

We return to our original scenario where Cardinal Player 1 is in Room 1, Player 2 in Room 2, and so forth. Suppose we ask the Cardinals to move in a different way. This time, Player 1 moves to Room 2, Player 2 moves to Room 4, Player 3 moves to Room 6, and so forth. So, for example, Cardinal Player 40211 moves to the room with a number that's double his jersey number—Room 80422 (*Figure 11.6*). Thus we've moved each Cardinal to a new room—the team now fills up all the even-numbered rooms. But all the odd-numbered rooms are now vacant. So we can now have Giant Player 1 check into Room

Fig. 11.6   *The entire team could be placed in just the even-numbered rooms.*

1, Giant Player 2 check into Room 3, Giant Player 3 check into Room 5, Giant Player 4 check into Room 7, and so forth (*Figure 11.7*). Thus we see that every Giant can have a room without sending any Cardinals out into the cold, and we've established a one-to-one pairing between the rooms and the players of both teams.

Fig. 11.7   *Every Giant can have a private room—the team will fill up the odd-numbered rooms.*

So doubling infinity does not produce a collection larger than the original infinity. Even when infinitely many new guests arrive at the already full Inn, the "No Vacancy" sign is still not required. Maybe there *is* always room at the Infinite Inn . . . or maybe not?

The question really is, Is *every* infinity the same size? That is, can every infinity be placed in a one-to-one pairing with the collection of counting numbers 1, 2, 3, . . . ? Our intuition and all the previous failed attempts to overfill the inn seem to indicate that infinity is infinity. But in fact, as we've already discovered, our intuition needs some retraining before we can accurately wrap our minds around the world of the infinite. To illustrate how our intuition can mislead us when considering the infinite, we leave the Infinite Inn and paddle our way to a plethora of Ping-Pong balls.

A PING-PONG PUZZLER—TOSSING BALLS OUT AT SPECTACULAR SPEEDS

Let's visualize infinitely many Ping-Pong balls lined up like little spherical soldiers. Each ball is labeled with a number—1, 2, 3, 4, 5, . . . , and so on forever—and the balls are lined up in numerical order. In our mind's eye there is no difficulty in visualizing such an impossible sight. Adjacent to this army of numbered Ping-Pong balls we see an enormous wooden barrel—larger than any barrel we've ever seen or ever will see. We're now ready to set off on a 60-second high-speed adventure that can be executed only in the realm of our imagination (*Figure 11.8*).

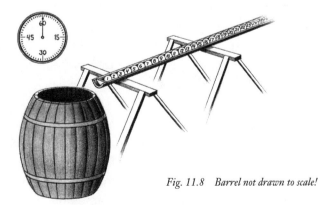

Fig. 11.8 Barrel not drawn to scale!

We start the timer, which has a red second hand that starts at 0 and makes one full rotation in 60 seconds. We begin leisurely: In the first $\frac{1}{2}$ minute (30 seconds), we drop the balls numbered 1 through 10 into the barrel, then reach into the barrel and remove and discard the ball labeled 1. This task can easily be accomplished in 30 seconds. No sweat.

Now we have 30 seconds left on the clock, and we speed things up a bit. In half the remaining time (15 seconds), we pour the next ten balls—numbered 11 through 20—into the barrel, then reach in and discard the ball labeled 2. You may think this is quick work, but you ain't seen nuttin' yet. In half of the remaining time ($7\frac{1}{2}$ seconds) we add the next ten balls in line—those numbered 21 through 30—and then we reach in the barrel and fish out and throw away Ball 3. We continue in this manner—in half the time remaining on the clock, we add the next ten numbered Ping-Pong balls into the barrel and then find and discard the ball in the barrel having the lowest number.

So let's see how this really works. In the first halftime (that is, the first 30 seconds: half of the 60 seconds), we toss Balls 1–10 into the barrel and remove Ball 1. In the second halftime (15 seconds, half of the time remaining), we add in Balls 11–20 and quickly remove Ball 2. In the third halftime ($7\frac{1}{2}$ seconds), we throw in Balls 21–30 and remove Ball 3. At the fourth halftime (a mere $3\frac{3}{4}$ seconds), we add in Balls 31–40 and remove Ball 4. This process continues until the timer reaches 60 seconds, at which time the experiment is over.

Notice that as we get closer and closer to the end, we find ourselves working at faster and faster speeds—remarkable speeds. In fact, in short order we would be working faster than the speed of sound—we wouldn't even be able to hear ourselves think. And soon after that we would be working faster than the speed of light—we'd probably just disappear. In fact, we'd soon be working far, far faster than the speed of light. Well, that's just impossible. True—but so is the sight of infinitely many numbered Ping-Pong balls. Let's leave the physical constraints of our real world behind and imagine doing this experiment in our minds.

The point of doing the rounds increasingly quickly is to enable us to envision doing infinitely many rounds of "10 in, 1 out" and then looking at the result. (Alternatively, we could do just one round each minute and then ask about what has happened after an *infinite* amount of time has passed. Either way is fine.)

In any case, after doing infinitely many rounds, the experiment and the 60 seconds are over. We're tired—working far faster than the speed of light can knock the wind out of even the heartiest of souls. Once we cool down and regain our composure, we walk over to the barrel and look inside. So what do we see? Is the barrel now filled with infinitely many Ping-Pong balls, does it contain a finite number of Ping-Pong balls, or is it empty? What do you think? Venture a guess.

A REASONABLE GUESS. One way to consider the question is first to consider how many balls are left in the barrel as we move through the experiment. After the first halftime we have 9 balls in the barrel (we placed 10 in and then removed one). After the second halftime we have 18 (we added 10 more and then removed another). After the third halftime we see 27 in the barrel. A pattern begins to emerge. After the fourth halftime, we see 36 (that is, 4×9); after the fifth halftime, we see 45 (5×9). Thus after each halftime the number of balls in the barrel has increased by 9. So a great guess is that there would be infinitely many balls left—since we will add 9 balls infinitely often. Sounds reasonable.

DOES THE BARREL HAVE BALLS? Let's suppose that someone believes that some balls are left in the barrel after the minute has expired. *Hmm* . . . Well, then that person can look into the barrel and see balls in there. *Interesting* . . . So that person can reach in and pull one out. *Fascinating* . . . Remember that each Ping-Pong ball has a number printed on its spherical surface. In particular, the ball that this person is holding has some counting number on it. We now ask this person to read off that number. We wait. And wait. What could that number be?

Could it be 4? No, because we know exactly when Ball 4 was removed: at the fourth halftime. Could it be 17? Well, no, since we removed Ball 17 at the seventeenth halftime. What about 1,009,328? Nope, that ball was removed at the 1,009,328th halftime. So what's the number on the ball pulled out from the barrel? Some might say "infinity," but "infinity" is not a number, so no ball has such a thing written on it. In fact, there is no number that could be written on this alleged ball—but all the balls have numbers, therefore it must be the case that there are no balls left. The barrel is utterly empty!

MAKING THE COUNTERINTUITIVE INTUITIVE. The incredible fact that there are no balls left in the barrel appears to run completely against our intuition, which illustrates the need to fine-tune our thinking so as to make the empty-barrel answer sensible.

To help us develop our intuition, let's modify the scenario just a touch. Suppose we start with *all* the numbered balls *in* the barrel and at each halftime we just remove the ball having the smallest number. So at the first halftime we remove Ball 1; at the second halftime we remove Ball 2; at the third halftime we remove Ball 3; and so forth for the infinitely many halftimes. In this modified scenario, clearly every ball is removed in turn, and thus after the 60 seconds (and infinitely many halftimes) the barrel is completely empty. In a way, this modified scenario shows that the putting in of ten balls at a time is just a red herring (or any other colored herring you like). The important point for us to focus on is the systematic removal of *all* the balls.

Our intuition on comparative sizes of collections—especially infinite ones—must be based solely on one-to-one pairings. Here again, we see a natural one-to-one pairing between the number on the ball and the number of the halftime. Ball 37 is paired up with the thirty-seventh halftime.

This shift in thinking requires some thoughtful reflection but is the key to our previous question—Can we ever run out of rooms at the Infinite Inn? Or, phrased in a more upscale mathematical manner, "Is there an infinity larger than the infinity of the collection of all counting numbers?" Is infinity always infinity, or do infinities, just like underwear, come in different sizes? Mystics, psychics, and even tailors will be of no help to us in confronting this incredible conundrum. All we have at our disposal is the basic idea that "same size" means "there's a one-to-one pairing."

BRINGING THE INTANGIBLE WITHIN OUR GRASP. Regardless of the details of these counterintuitive conundrums, one thing about infinity is certain: We can now deal with infinity in a much more straightforward manner. We started with a vague feeling about something incomprehensible and brought it down to a concrete idea. We found the key to grappling with infinity when we focused on the simple and familiar idea of 5. It was child's play to match up fingers, but that was the critical move. The consequences of exploring the simple idea of one-to-one pairing are far from exhausted, but already we see the power of resisting the urge to bask in a fog of vagueness and, instead, focus on understanding the simple and familiar deeply.

CHAPTER 12

IN SEARCH OF SOMETHING STILL LARGER

A Journey Beyond Infinity

There is no smallest among the small and no largest among the large;
but always something still smaller and something still larger.

—Anaxagoras

Obviously . . . We need not call our CPA when a child innocently challenges us to a "battle of the biggest." We say, "Ten," and with great pride the kid responds, "Eleven." After we're egged on to try again, we say, "One thousand seventeen," and the little angel, not the least nonplused, shoots back "One thousand eighteen!" Our headache builds as we keep pace with our budding numerologist's enthusiasm, and finally we decide to bring these festivities to a close. We say, "Infinity," to which our young opponent gleefully counters with "Infinity plus one!" With an embarrassing degree of delight we say, "Ha!" and explain that when we add one to infinity, infinity doesn't budge an inch from its original size. In fact, even if we add infinity to itself, still we see no increase in largeness. Infinity, our intuition tells us, comes in just one size: XXX. . .-Large. Game over.

Surprise . . . If that child were sufficiently precocious, the game *would* continue—in fact, it would continue *forever*. For infinity, just like numbers, comes in an ever-growing tower of sizes. One infinity does not fit all!

————

A SEARCH FOR SOMETHING BEYOND INFINITY

You'd think that once we got to infinity, we'd have seen it all, and looking for something bigger would be not only futile but downright greedy. When we thought about infinity as an incomprehensible, all-inclusive everything, it made no sense to compare things that were infinite—after all, infinity was everything. When we vaguely viewed infinity as "the ultimate in bigness," it made no sense to ask whether there was something even larger.

But now we are developing a more familiar—dare we say intimate?—relationship with infinity. In our minds we are seeing a less fuzzy vision of infinite collections. We've juggled infinitely many Ping-Pong balls without missing a beat and managed hotels with infinitely many rooms without fear of stolen shampoo. We have compared infinite collections and lived to talk about it. Perhaps we can now face the seemingly impossible question: Is there something bigger than infinity, or are all infinite collections the same size?

A FOCUS ON ONE-TO-ONE PAIRINGS

Asking whether all infinities have the same size—or, mathematically speaking, whether they have the same cardinality—really means, "Suppose someone gives us two infinite collections. Is it always possible to pair them up in a one-to-one manner?"

So far, our encounters with the infinite have resulted in successful one-to-one pairings with the collection of counting numbers. However, now we will describe a collection that is *larger* than the collection of counting numbers—that is, we will see that it is impossible to pair up the elements of this new collection with the counting numbers in a one-to-one way—we'll always run out of counting numbers first. Therefore, this new collection is actually bigger than the infinity of all counting numbers!

In order to describe this new collection for which a one-to-one pairing with the counting numbers is impossible, we introduce a little game that is easy to play and easy to figure out, yet has enormous implications—in fact, infinite implications.

DODGE BALL

Dodge Ball is a board game for two players, whom we imaginatively call Player One and Player Two. The two players have different boards (*Figure 12.1*) on which they play alternately for six thrilling

DODGE BALL — **the game**

Player One's game board

Player Two's game board

Fig. 12.1

turns each. During the whole game both boards are always visible to both players.

After doing appropriate warm-up exercises, Player One starts the game by writing an X or an O in each of the six boxes of his first horizontal row. Then Player Two ponders the situation and fills in the first square of her row with an X or an O (*Figure 12.1a*).

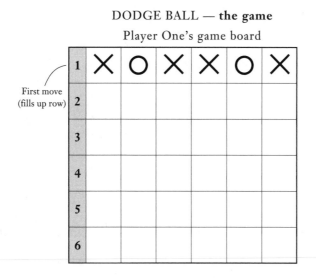

DODGE BALL — **the game**

Player One's game board

Player Two's game board

Fig. 12.1a

Now Player ② puts one X or O.

Player One scratches his head and then takes his second turn, which consists of filling in his second row with any sequence of X's and O's that he decides on. Player Two looks carefully at Player One's move and responds by writing an X or an O in the second box of her board (*Figure 12.1b*). Player One and Player Two continue taking alternate turns, Player One always writing a whole row of X's and O's and Player Two filling in just one box (*Figure 12.1c–e*). They take six turns each, after which time Player One has written down six

Fig. 12.1b–e

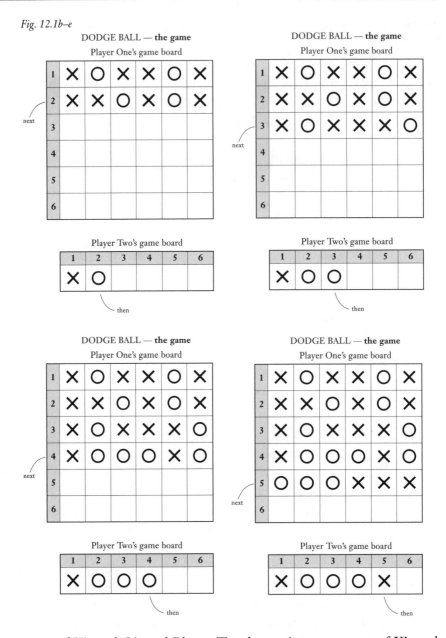

rows of X's and O's and Player Two has written one row of X's and O's (*Figure 12.1f*).

But who wins? Player One wins if any one of his six rows has the same sequence of X's and O's as Player Two's single row of X's and O's; that is, Player One wins if one of his horizontal rows *matches*

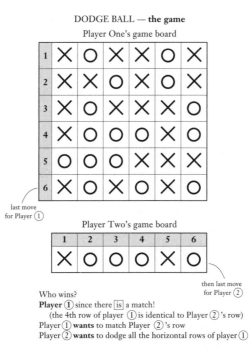

DODGE BALL — **the game**

Player One's game board

1	X	O	X	X	O	X
2	X	X	O	X	O	X
3	X	O	X	X	X	O
4	X	O	O	O	X	O
5	O	O	O	X	X	X
6	X	O	X	O	X	O

last move
for Player ①

Player Two's game board

1	2	3	4	5	6
X	O	O	O	X	O

then last move
for Player ②

Who wins?
Player ① since there is a match!
   (the 4th row of player ① is identical to Player ②'s row)
Player ① **wants** to match Player ②'s row
Player ② **wants** to dodge all the horizontal rows of player ①

Player Two's row. Only horizontal rows count for a match—up and down or diagonal rows don't play a role in this game. Player Two wins if her row is different from each of Player One's six rows, that is, she wins if she has *dodged* each of Player One's six rows.

Some people have called this a bored game, because we don't have to consult Bobby Fischer to plumb its subtleties. But Dodge Ball, though far simpler than chess, actually goes much further because Dodge Ball will take us to different sizes of infinity.

HOW PLAYER TWO CAN ALWAYS BE A WINNER

It doesn't take long to realize that Player Two has a winning strategy: She simply does the opposite of whatever Player One has just done. That is, after Player One has written down his first row, Player Two looks at Player One's first square and then in her own first square puts an X if Player One used an O and puts an O if Player One used an X. By doing so, Player Two has guaranteed that her row will not be identical to Player One's first row—the rows will differ in at least the first square. Then Player One writes down a second row, and Player Two looks at the second square of Player One's second row. Once again, Player Two puts the opposite mark in her second square. Player Two continues to follow this strategy and wins.

This game is simple and far from exciting—alas, Parker Brothers is not going to dash out and market Dodge Ball in place of Monopoly. For our purposes, however, its simplicity is a virtue. And the game does have a noteworthy feature: Player Two's winning strategy would work not only on a 6 × 6 board but on a square board of any size. Would there be any change in strategy with an 8 × 8 board (*Figure 12.2*)? No, the strategy is the same and Player Two wins yet again. But what does this ridiculous game have to do with different sizes of infinity?

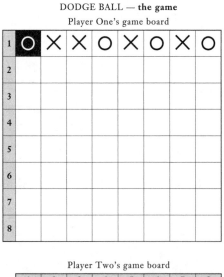

Fig. 12.2   Player ② *uses the winning strategy at the first round of play.*

INFINITE DODGE BALL

We're now ready to pole-vault from the obvious up over to the abstract and ask what it would mean to play Dodge Ball on an *infinite* board. That is, the board has infinitely many rows—Row 1, Row 2, Row 3, . . . and so on forever, one row for each counting number—with each row containing infinitely many squares (*Figure 12.3*).

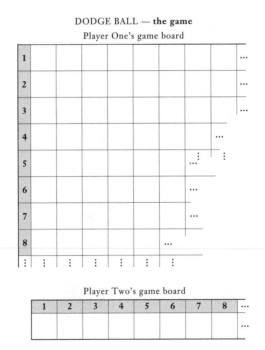

DODGE BALL — **the game**

Player One's game board

Player Two's game board

Fig. 12.3 *Infinite Dodge Ball game board*

We could imagine this Dodge Ball game being played forever—Player One making an infinite row of X's and O's, then Player Two making just one mark, then Player One filling in the second, infinitely long row, then Player Two making a second mark . . . and so on *forever*.

Let's see what would happen if Player Two used the same strategy that she used for the 6 × 6 game. If we waited to the end of all these infinite number of moves, it isn't too difficult to see that Player Two would have created an infinitely long row that differs from every one of Player One's rows. Specifically, Player Two's row would definitely not be the same as Player One's *first* row, because the entry in the *first* square of Player Two's row was chosen to be different from the entry in the *first* square of Player One's *first* row. Likewise, Player Two's row would definitely not be the same as Player One's *second* row, because the *second* square of Player Two's row is different from the *second* square of Player One's *second* row. Row by row, we know that Player Two's row differs from each of Player One's rows. So Player Two has dodged all of Player One's infinitely many rows (*Figure 12.4*).

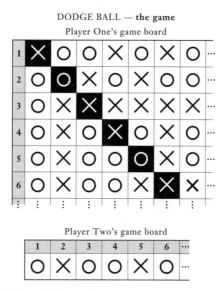

Fig. 12.4   Player ② *dodges and wins.*

Of course, some might say that the game would never end, because the players have infinitely many turns to take. But suppose that, as was the case with the Ping-Pong ball conundrum, we simply speeded up the game, making it so incredibly fast that all of the infinite number of turns could be completed in 60 seconds. Then the game would end in a minute, and Player Two would win, eager to play again, this time perhaps for money.

This game seems straightforward, and yet its implications are surprisingly profound and counterintuitive. Let's think about a slight modification of the game. Suppose Player One makes *all* of his moves before Player Two makes any of her moves. That is, Player One takes his whole game board and fills it in completely. Then Player Two could look at all the rows and try to write down a row that is clearly different from each of the rows that Player One has written down. Of course, Player Two's winning strategy of focusing on the diagonal entries in Player One's board—the first square of the first row, the second square of the second row, and so on—and writing the opposite still works fine, and she wins as before. This game sounds even more unfair to Player One, since he has to commit to the whole board before Player Two has made any commitment on her row. At this point, Player One is considering changing his last name to Two.

A POTENTIAL WINNING MOVE FOR PLAYER ONE

Let's put ourselves in Player One's poor shoes. Player One is desperately trying to write down rows so that one of them will be exactly the same as the row that Player Two is going to write later.

Player One might think of the following intriguing strategic idea: "One way to make certain that Player Two duplicates a row of mine is for me to simply write down *all* possible rows of X's and O's. That way, no matter what Player Two writes down, she is certain to duplicate one of my rows." In other words, it is like spending millions of dollars and buying a lottery ticket for each possible set of numbers that could be drawn. Then we are certain to win, no matter which numbers bubble up.

On the surface, this scheme appears to be a great strategy for Player One to consider. After all, Player One has infinitely many rows at his disposal. So can Player One use this strategy and, at last, declare victory?

Suppose it were possible for Player One to write down every possible run of X's and O's—one in Row 1, one in Row 2, one in Row 3, and so forth. If all the possible strings of X's and O's *could* be listed in such a manner, then Player Two would be in trouble.

WHY THE EXHAUSTIVE STRATEGY IS NOT THAT EXHAUSTING

Player Two is not in trouble. No matter what Player One writes down, Player Two *always* has a response. Player Two can always write down a row of X's and O's that differs from each and every row that Player One wrote. We know that Player Two has a simple way to dodge all those rows: She just looks at Player One's board filled in with all the X's and O's, and then she goes down the diagonal and switches each entry to make her row. So Player Two's winning strategy really proves that it is *impossible* for Player One to successfully use his exhaustive strategy. In other words, the endless game board is not long enough for Player One to list every single possible row of X's and O's.

What is Player One doing when he completely fills in an infinite game board? Player One is pairing each row number, 1, 2, 3, 4, 5, . . . , with an infinite list of X's and O's. So each counting number is associated with an infinite run of X's and O's.

INFINITY COMES IN MORE THAN ONE SIZE

Player Two's winning strategy implies that there is no way to make a one-to-one pairing between the row numbers and all possible infinitely long rows of X's and O's. In other words, the collection of all possible rows of X's and O's is *larger* than the collection of counting numbers, which are the row numbers. This diagonal strategy demonstrates that the size of the collection of all possible infinite runs of X's and O's is larger than the size of the counting numbers (the collection of row numbers).

This realization is so astounding that we explain it again in slightly different words. It is astounding because we have conclusively shown that there are more possible infinite runs of X's and O's than there are counting numbers. We accomplished this by proving that it is *impossible* to make a one-to-one pairing between the counting numbers and the collection of all possible infinitely long rows of

X's and O's—for every attempt offered, we can march down the diagonal and produce a run of X's and O's that is definitely not paired up with any number.

So, not all infinite collections can be put in one-to-one correspondence with each other. In particular, here are two different sizes of infinity: the size of the collection of counting numbers and the size of the collection of all possible endless rows of X's and O's. Infinity does not come in just one size. To put it mathematically, we have shown that the cardinality of the collection of counting numbers is not the same as the cardinality of the collection of all possible endless rows of X's and O's. Our reaction: "Wow!"

PAUSE FOR BREATH

We considered just leaving a blank space here, because we hope that you will feel the need to pause and let the implications of infinite Dodge Ball sink in. No individual step in the reasoning is difficult, but the consequence is an enormous pill to swallow. Somehow we started with a silly little game with an easy strategy, and a few paragraphs later, POW!—we've shot past infinity.

Let us assure you that the reasoning is ironclad and within the reach of every reader, but the result is not one that is easy to accept. In the late 1800s when Georg Cantor first proved that infinity comes in more than one size, the mathematics community was deeply skeptical and put Cantor in the middle of a raging controversy—math wars are not a new invention. The arguments were so vituperative and personal that poor Cantor ended up in a lunatic asylum. Now, fortunately, we can embrace different sizes of infinity with our wits pretty much intact. But the result is so striking that it deserves at least a couple of different views.

To find another collection so massively infinite that a one-to-one pairing with the counting numbers is impossible, we throw caution to the wind and sheets into the laundry as we face, head on, the dirty issue of cleaning our Infinite Inn.

CARDINALITY CLEANERS

We now return to the Infinite Inn to discover one of the many challenges and surprises in the world of infinite hostelry. Remember that our infinite hotel has rooms numbered 1, 2, 3, 4, 5, . . . Now, of course, with all those rooms there is a heck of a lot of room cleaning to be done. So this hotel contracts with Cardinality Cleaners, which specializes in cleaning major messes. But naturally the hotel isn't full every night, so each morning the hotel manager calls Cardinality Cleaners and tells the manager which rooms need to be cleaned. One morning, perhaps only the even-numbered rooms need cleaning; on another morning, only rooms 1, 2, 3, 17, and 307 need maid service.

Well, as you can imagine, cleaning any Infinite Inn requires highly specialized and trained personnel, and Cardinality Cleaners has got them all on call. In fact, each of the employees specializes in only one particular collection of rooms. There is one person who is called in when all the even-numbered rooms need cleaning. There is a different person who does the job when all the even-numbered rooms *except* 2 and 12 need cleaning. There is yet another person who is called only when Rooms 2, 4, 8, 16, 32, 64, . . . need cleaning. A different person is brought in to clean the collection of Rooms 1, 2, 6, 1007, and 20149—and no others. The worst job is that of the poor soul who is called in when *every* room needs to be cleaned, and the cushiest job goes to the lucky specialist who is called in when *no* rooms need cleaning—nice work if you can get it.

In other words, given any particular list of dirty rooms, there is exactly one person at Cardinality Cleaners who is called in to clean precisely those rooms—no more and no less. So only one person works on any one day, and every possible configuration of dirty rooms is covered by a different employee at Cardinality Cleaners. Seems like a funny way to run a business. You might think that it would be better for the hotel to assign one room to each person and have that person clean only that room. But whether that would be better or not, for this fictitious scenario we will have to accept the ridiculous business plan employed by Cardinality Cleaners—each day only one person cleans.

HOLIDAY CHEER

Despite some amazing inequities in workload (the person who cleans only Rooms 23, 48, 102, and 100034567 tends to avoid making eye contact with the cleaner whose job it is to clean all the odd-numbered rooms), the crew at Cardinality Cleaners is a happy one. The goodwill between management and cleaners finds no higher expression than during the December holiday season. As an end-of-year bonus to all cleaners, the owners of Cardinality Cleaners book every room at the Infinite Inn, and every cleaner is offered a weekend stay at the Inn. This generous offer pushes the envelope of hotel hospitality—each cleaner is to get a private room. They all arrive in good holiday cheer (eggnog is served before they all head to the hotel) and ask the desk clerk to check them in. At this point, beads of sweat begin to appear on the clerk's brow. Where will each cleaner go?

Some cleaner will be given Room 1, another will be placed in Room 2, another in Room 3, and so on. Remember that there are no bunkmates—each specialized cleaner was promised (and certainly deserves) a private room. So the clerk tries to find one room for each cleaner, but every assignment he attempts seems to fill up the hotel before everyone has his or her own room. Being at an inn with infinitely many rooms, the clerk is accustomed to being able to accommodate all his guests. But today he is stymied—every assignment of cleaners to rooms seems to leave some cleaners roomless. Why is that happening? Let's back up and get to know a little more about our cleaners.

PLAYING DODGE BALL WITH THE CLEANERS

Of course, every cleaner has a name, and the management at Cardinality Cleaners knows them all. More typical for corporate America, however, would be for each employee to have a code for identification. In fact, one coding scheme would be to name each person by the collection of rooms he or she cleans. We could do this by assigning each person an endless run of X's and O's. Each letter would represent a hotel room, in order—the first letter is for Room 1, the second letter for Room 2, and so on. The letter is an X if the cleaner cleans that room and an O if the cleaner does not clean that room. So, for example, Cleaner OXOXOXOXOXOXOXOXOX. . . is the person who cleans all the even rooms (notice that the X's are in the even-placed spots), while Cleaner XXOXOOXOOOOOOOOO. . . is the person who cleans only Rooms 1, 2, 4, and 7.

Now suppose we've filled all the rooms at the Infinite Inn with cleaners. Who's where? Let's list all the room numbers down the page, and next to each room number let's write the string of X's and O's that identifies the cleaner staying in that room. Now, using the diagonal-based strategy that Player Two used to win at infinite Dodge Ball, we can name a cleaner who is not in any room. We could just switch the X's and O's along the diagonal, as in Dodge Ball, and come up with the name of another cleaner—one who is different from all the cleaners who have rooms. *This* cleaner is on the streets; you might call this poor soul a "street cleaner" (*Figure 12.5*).

Fig. 12.5

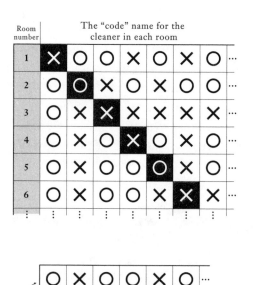

The "name" of the cleaner that definitely is without a room

The Dodge Ball–winning strategy demonstrates that no matter how the desk clerk tries to distribute the rooms, we can always name a roomless cleaner. Thus a one-to-one pairing between the two collections is impossible, and we discover that the collection of all the employees at Cardinality Cleaners is actually *larger* than the collection of counting numbers (room numbers). Every attempt that the desk clerk might make to assign cleaners to rooms will always fail. For any possible assignment, we can identify at least one cleaner who is left out in the cold. We are led to the inexorable conclusion that there are more employees of Cardinality Cleaners than there are rooms at the Infinite Inn.

To put it more formally, there is no one-to-one pairing between the collection of counting numbers (the room numbers) and the collection of all possible collections of the counting numbers (all possible collections of rooms that may need to be cleaned on any particular morning).

MORE THAN ONE SIZE OF INFINITY YET AGAIN

Again we are faced with the reality that infinity comes in more than one size. In particular, the collection of employees of Cardinality Cleaners is definitely larger than the collection of rooms at the Infinite Inn, which is already infinite. It is impossible to create a one-to-one pairing between the collections. Some infinities are bigger than others.

Some may worry that Cardinality Cleaners has somehow swept something under the rug. We hasten to assure these people that this is not the case. Our little tale does cleanly and conclusively prove that infinity comes in more than one size. The realization that there are different sizes of infinity is dramatic and counterintuitive, and requires very sober consideration. (Infinity will not come into focus with a glass of Merlot—although feel free to try to prove us wrong.)

NUMBERS ON THE LINE

While it's a stretch to dream up "real-world" infinite capers galore, infinity is natural and there for the taking in the math world. In fact, all we have to do is hark back to the number lines that graced our classrooms and perhaps our desks in elementary school. Those readers who would rather not relive those childhood math memories (or nightmares) are encouraged to ignore this section and move on.

All the numbers on a number line can be expressed as decimals and are called *real numbers*. Let's just focus our attention on the decimal numbers between 0 and 1. Each decimal number between 0 and 1 starts off with a 0 and then a decimal point, followed by an infinitely long string of digits such as 0.500000. . . or 0.001237733. . . or 0.12345678910111213141516. . . . Each infinitely long string of digits represents a single real number between 0 and 1—a single point on the number line.

Is there a one-to-one pairing between the decimal numbers and the counting numbers? That is, suppose we have a barrel of Ping-Pong balls, each labeled with a counting number (1, 2, 3, 4, . . .), and we have a barrel of golf balls, each labeled with a decimal number (such as 0.3456351. . . or 0.123123144. . .). Our question is, "Can we make a one-to-one correspondence between the Ping-Pong balls and the golf balls?"

The answer is no. The reason, yet again, is the Dodge Ball–winning strategy. Suppose every Ping-Pong ball *could* be paired up with a golf ball. So Ping-Pong Ball 1 would have some decimal-labeled golf ball associated with it, Ping-Pong Ball 2 would have another golf ball associated with it, and so on. We could write the pairing down in a chart where the names of the Ping-Pong balls are in the left-hand column and the labels of the golf balls are in the right column (*Figure 12.6*).

If a one-to-one pairing existed between these two collections, then every golf ball would be used up. But that is not possible, in view of the Dodge Ball strategy. If someone proposes a pairing, we can always find a golf ball that has not been paired up. Which one? The one whose decimal number is determined by looking down the diagonal of the list and changing each decimal digit to create the name of a ball that cannot be on the list. So our reasoning actually shows that there are more decimal numbers—numbers on the number line—than there are counting numbers. Who would have guessed this amazing fact back in elementary school, where the number line's primary use was for spitball target practice?

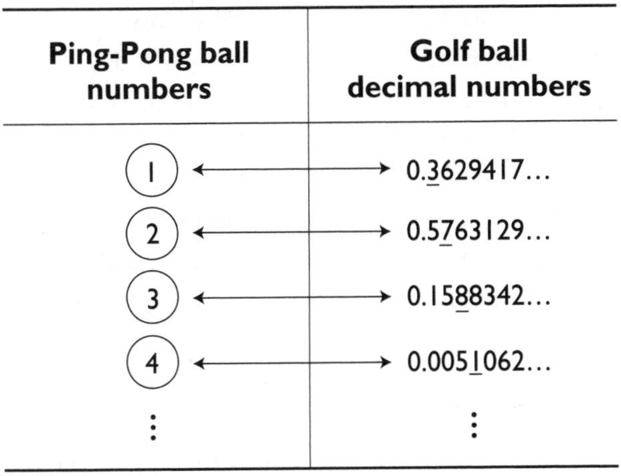

Ping-Pong ball numbers	Golf ball decimal numbers
① ←—————→	0.3629417...
② ←—————→	0.5763129...
③ ←—————→	0.1588342...
④ ←—————→	0.0051062...
⋮	⋮

Fig. 12.6   An attempt at a one-to-one pairing

INFINITIES COME IN INFINITELY MANY SIZES

Once we discover that infinity comes in different sizes, we are led to ask, "Is there more?" We have found infinities that cannot be put in one-to-one pairing with the counting numbers—for example, all the employees of Cardinality Cleaners. But can we find infinite collections that are bigger still? Can we find an infinity that is even larger than the cleaning crew of Cardinality Cleaners?

Yes, indeed we can. When the employees of Cardinality Cleaners finally realize that the Infinite Inn is too small for them, they all pack into a bus (a mighty large bus, as you might have guessed), drive down the road, and find an even larger hotel, the Grand Hotel Cardinality. This beautiful new hotel made of steel and glass has exactly one room for each member of the incredibly large cleaning crew of Cardinality Cleaners. They enjoy their stay at this magnificent hotel. By the way, the room numbers are all decimal numbers, since there are more rooms than there are counting numbers—so of course they don't use room service, since it would take forever to say their room number: "Room service? I'd like to place an order. I'm in room

0.54233389373659722290576506744485958746845859595..."

They'd starve long before they ever come close to revealing their precise location.

Being curious about cleaning, the folks from Cardinality Cleaners ask the manager how the rooms at the Grand Hotel Cardinality are cleaned. The manager explains that the method is the same as for the Infinite Inn—he outsources the work. He uses a cleaning company called Uncountable Cleaners. Every morning he just calls up and says which rooms need to be cleaned (a huge job in itself), and Uncountable Cleaners sends over one amazingly dedicated cleaning person, the person whose specialty is cleaning exactly that collection of rooms.

You see where this is going, right? Using the exact same reasoning as we did earlier for the Cardinality Cleaners trying to enjoy their all-expenses-paid weekend at the Infinite Inn, we discover that there are more cleaners at Uncountable Cleaners than there are rooms at the Grand Hotel Cardinality. In this fashion we find that there are infinitely many hotels of ever-increasing size, with their cleaning crews always larger than the hotels that they clean. This remarkable insight leads us to the realization that there are *infinitely* many sizes of infinity.

Could there be a largest infinity? Could we finally arrive at the mother of all infinities, the infinity that we imagined in the first place? An infinity that really is all-encompassing?

Well, no. Why? Because a hotel with as many rooms as this alleged mother of all infinities needs to be cleaned, and that cleaning crew—one member of the cleaning crew for each collection of rooms that might need to be cleaned—is too large an infinity to spend the weekend in that hotel. Every infinity has a superior.

A THINKING STRATEGY THAT WORKS

We were able to understand infinitely many infinities by starting with a simple idea and following it mercilessly and open-mindedly as far as we could. The habit that worked here and is powerful for creating new ideas anywhere is: Explore the consequences of a new idea.

In this case, we zeroed in on the idea of comparing collections by making one-to-one pairings, and then we followed the consequences of that new perspective. We explored various collections that could be paired one-to-one—for example, the rooms at the Infinite Inn paired with what seemed like a collection twice the size, the two infinite baseball teams. Then we found collections that could not be paired, such as the collection of all possible infinite lists of X's and O's in Dodge Ball that could not be paired with the counting numbers 1, 2, 3, 4, ... These results seemed counterintuitive at first and were hard to swallow, despite being correct. It is hard to accept results of reasoning when the results contradict what we first guess. Perhaps the biggest challenge here was one that confronts us frequently in everyday life—the challenge of keeping an open mind.

CLOSING THOUGHTS

Mathematics is a liberating entertainment. We can discard the kid gloves of reality's restrictions and let our minds play where they will. Mathematics is about what can be thought, what can be imagined, what can be dreamed. And it leads us to new truths. Exploring the deep consequences of simple ideas takes us on a journey of startling sights and unexpected insights. We fold paper to find patterns, we count spirals to find numbers, we unfold cubes of all dimensions to see four-dimensional worlds where we cannot physically roam but where our minds can romp freely.

Mathematics is about casting off the constricting coil of bound thought. The mathematical world is full of wonder, but our guide is the clear principle of following simple ideas to logical conclusions. We can embrace this basic strategy of mathematical thinking to guide us in our everyday lives as well. We have seen here just a tiny glimpse of the vast richness that our minds can create. Mathematics and our imaginations have no bounds, no ends, no finish line. Every horizon reached opens new horizons more glorious still.

ACKNOWLEDGMENTS

First and foremost, we wish to express our deepest appreciation to three people who inspired us with their constant enthusiasm throughout this project and made this book a reality. They are Lisa Queen from IMG, and, from W. W. Norton & Co., Maria Guarnaschelli and Erik Johnson. Lisa, Maria, and Erik energized us with their excitement for our vision and, through their insightful remarks, considerably improved the final product.

Also from Norton we wish to thank Eleen Cheung, Julia Druskin, Starling Lawrence, Jeannie Luciano, Drake McFeely, Bill Rusin, Erin Sinesky, and Nancy Palmquist for their creative and artful talents and for their support of our project. Beyond Norton, we thank Katya Rice for her outstanding copy editing; Alan Witschonke for his beautiful original illustrations; Jamie Keenan for the terrific dust jacket design; Soonyoung Kwon for her clean design, Jamie Kingsbery for rendering the fractal images; and Charles Radin for the Pinwheel Tiling image.

Finally, we wish to express our gratitude to all the family members, friends, colleagues, and students who have been a constant, joyful source of inspiration and encouragement over many years. Mike especially thanks his wife, Roberta, and children, Talley and Bryn.

FURTHER RESOURCES

The Heart of Mathematics: An Invitation to Effective Thinking, E. B. Burger and M. Starbird, Key College Publishing, 2004.

The Joy of Thinking: The Beauty and Power of Classical Mathematical Ideas (a video course), E. B. Burger and M. Starbird, The Teaching Company, 2003.

The publisher and authors make grateful acknowledgment for permission to reproduce the following material:

Note: Page numbers in *italics* refer to illustrations and captions

aesthetics, 116, 121–45
  of ancient Greek eyecups, 126–27, *127*
  in architecture, 135–36, *135–36*
  destruction of volume concept and, 128–29
  in modern art, 128–29, *129*
  of Parthenon, 125–26, *126*, 128
  in portrait of Saint Jerome, 127–28, *127*
  see also Golden Ratio; Golden Rectangle;
    Golden Triangle
AIDS (acquired immune deficiency syn-
    drome), 58
air safety example, 56–58
Allen, Paul, 48
Allen, Woody, 233
Anaxagoras, 246
Antony, Mark, 67
architecture:
  aesthetics in, 135–36, *135–36*
  Golden Rectangle in, ix
area paradox puzzle, 100–101, *100–101*,
    111–13, *112–13*
Aristotle, 123
art:
  fourth dimension in, 227–29, *227–29*
  modern, aesthetics in, 128–29, *129*
  see also aesthetics; architecture
average income example, 47–48

Bacon, Kevin, 88
bell-shaped curve, 49–51, *49–51*
Benny, Jack, 79
bias, 45–47
Bible, 79
Bigallo, *see* Fibonacci
Big Bang, 84
billions, 84–85

birthdays, 17–19
Boethius, 3
Bonacci family, 108–9
Booth, John Wilkes, 4
bouncing ball example, 37–40, *38–39*
Brahe, Tycho, 83
Brutus, Marcus Junius, 66
butterfly effect, 21–22, 23

Caesar, Julius, 66–67
Caesar Cipher, 66–67
calculus, 74
California, 84
Cantor, George, 257
cardinality, 239, 247, 257
  see also one-to-one pairings
center of gravity, 93–96, *93–96*
chaos, ix, 1–2, 20–41
  accident and, 31
  in actual world, 23
  in bouncing ball example, 37–40, *38–39*
  butterfly effect and, 21–22, 23
  computational, 24–31
  defined, 2, 23
  in double pendulum example, 33–34,
    33–35
  in drip patterns, 34–36
  as fundamental feature of nature, 40
  magnets and, 36–37, *36–37*
  mathematical, 2, 23
  in moving objects, 32–35, *33–35*
  in physical systems, 32–35, *33–35*
  predictions and, 23, 32, 41
  and sensitivity to initial conditions, 37–41
  squaring example of, 24–31
codes, 66–67

codes (*continued*)
  enigma, 155
  public key, 69–73, 75, 77
  see also cryptography
coincidence, ix, 3–19
  of birthdays, 17–19
  dealing with, 4–5
  intuition and, 5, 19
  karma and, 11–12
  Lincoln-Kennedy parallels and, 4–7
  lotteries and, 16
  magic and, 19
  minutiae and, 5–6
  monkeys metaphor and, 10–11
  predictions and, 12–15
  randomness and, 10–11
  twin studies and, 7–9
Cold War, 65
computers, computing, 147
  finite-state automata and, 155–57, *156–57*
  halting problem in, 157
  programs, 156–57, *156*
Congress, U.S., 4, 88
continued fraction, 113–14
Copperfield, David, 201–2
counting:
  exact, 79
  to infinity, 232–33
  order-of-magnitude, 79–80
  see also numbers
Crichton, Michael, 160–62
Crucifixion, Corpus Hypercubicus, The,
      (Dali), 227, *227*
cryptography, vii, ix, 65–77
  Caesar Cipher and, 66–67
  in coding-decoding process, 70–72
  factoring and, 67–69
  public encryption scheme and, 65, 69–72
  public key encryption and, 69–73, 75, 77
  security and, 72
  sending secrets and, 66–69
  spies and, 65–66
  value of information and, 72–73
Cryptoquote, 67
cube(s):
  five-dimensional, *224*
  four-dimensional, 222–26, 222–27,
      222–24, 225–27
  models, 195–97, *195–96*
Cummings, E. E., 201

Dali, Salvador, 227
da Vinci, Leonardo, 127
Debussy, Claude, 123, 129–30, *131*
decimal numbers, 261–63, *263*
Declaration of Independence, 89

de Morgan, Augustus, 166
deoxyribonucleic acid, *see* DNA
de Pisa, Leonardo, *see* Fibonacci
Depression, Great, 44
destruction of volume, concept of, 128–29
disappearing rabbit example, 212–14, *212–14*
Disraeli, Benjamin, 42
Divina Proportione, De (da Vinci), 127
DNA (deoxyribonucleic acid), 122, 180–84,
      182
  knots and, 181
  replication of, 184
  storing, 183
  supercoiling of, 183–84, *183–84*
dodge ball, 248–52, *251–52*, 255–56, 260
double pendulum, 33–34, *33–35*
doughnuts:
  topology and, 174–77, *175–77*
  see also Klein bottle; Möbius band
Dragon Curve fractal, ix, 122, 146–47, *147*,
      163–65, *163*
  Golden Triangle and, 164–65
drip patterns, 34–36
Duchamp, Marcel, 227–28

Earth:
  age of, 79–80
  population of, 85
education, 47–48
  see also SAT
Egypt, ancient, 81–82
elasticized world, *see* topology
election of 1936, 43–44
Electoral College, 44
ELISA blood-screening test, 58
encryption, public key, 65, 69–73, 77
  prime numbers and, 71–72, 75
enigma code, 155
Enron, 22
exact counting, 79
Expressionism, 128
eyecup design, 126–27, *127*

factoring, 67–69
Fibonacci (Leonardo de Pisa), 108–9
Fibonacci numbers, 108–13, 139
  area paradox puzzle and, 100–101,
      100–101, 111–13, *112–13*
  Golden Ratio and, 116–17, 119, 124–25
  in human form, 139–40, *139–40*
  Lucas numbers and, 116–19, *117–18*
  in music, 130
  rabbit-generation question and, 109–10,
      110
  ratio of adjacent, 110–11, 113–16, *113–15*
  in spiral patterns, 108–9

financial predictions, 12–15, 22
finite-state automata, 155–56, *156–57*
Fischer, Bobby, 251
five-dimensional cube, *224*
four-dimensional cube, 222–26, *222–27*
    building, 222–24, *222–24*
    visualizing and rendering, 225–26, *225–27*
fourth dimension, 201–30
    allure of, 229–30
    in art, 227–29, *227–29*
    boundaries and, 209–12, *210–11*
    building cube of, *see* four-dimensional cube
    dimension as physical freedom and, 202–4,
        215
    disappearing rabbit example and, 212–14,
        212–14
    holeless Klein bottle and, 220–21, *221*
    knotted rope problem and, 215–19, *215–19*
    as stacked three-dimensional spaces,
        207–9, *208*
    time as model of, 228–29
    "What if?" question and, 199
fractals, 121
    defined, 163
    Dragon Curve as, 163–65, *163*
fractions:
    defining trait of, 114
    ones over ones 113–16, *113–15*
France, 128
future, predictions of, 23, 32, 41

Gallup, George, 44–45
Gallup Poll, 45
Gates, Bill, 48, 79, 84–85, 232
Gateway Arch, 167
genetics, personality and, *see* twin studies
gigabyte, 84
Goldbach, Christian, 75
Goldbach Conjecture, 75–76
Golden Ratio, 116, 128, 129, 133, 140
    Fibonacci numbers and, 116–17, 119,
        124–25
    Golden Triangle and, 124
    in Greek eyecups, 126–27, *127*
    Lucas sequence and, 116–19, *117–18*
    in music, 129–30, *131*
    starting seeds and, 116–17, 119
Golden Rectangle, ix
    in architecture, *130*
    construction process of, 131–33, *132–33*
    defined, 125
    in Greek eyecups, 126–27, *127*
    human form and, 139–40, *139–40*
    logarithmic spiral within, 138, *138*
    in modern art, 128–29, *129*
    in Parthenon, 125–26, *126*, 128

in portrait of Saint Jerome, 127–28, *127*
    proportions of, 124–25, 127–28, *127–28*,
        135, *135*
    proving, 133–35, *134*
    smaller Golden Rectangles within, 136–38,
        136, 138
    see also Golden Ratio; Golden Triangle
Golden Triangle, 141–45, *141–42*
    Dragon Curve fractal and, 164–65
    Golden Ratio and, 124
    Pinwheel Tiling and, 143–45, *144*
    regeneration process and, 142–43, *143*
    translational symmetry of, 143, *143*
Great Depression, 44
Greeks, ancient, 125, *131–32*
    eyecup design of, 126–27, *127*
    see also Parthenon

halting problem, 157
Hamlet (Shakespeare), 10–11
Heraclitus, 100
HIV (human immunodeficiency virus),
    58–60, 89
Houdini, Harry, 201

identification diagram, 187–89
Impressionism, 128
infinity, vii, 231–67
    counting to, 232–33
    dodge ball strategy and, 248–52, *251–52*,
        255–56, 260
    doubling, 239–41, *240*
    half of, 238–39, *238*
    one-to-one pairings and, *see* one-to-one
        pairings
    pairing means equal concept and, 235
    Ping-Pong balls example and, 231, 241–44,
        241
    plus one, 236–37, *236*
    sizes of, 241, 244, 261, 264
    unknown numbers and, 232–33
    "What's next?" question and, 199–200
Interior of the Fourth Dimension (Weber), 229,
    229
Internet, 65, 66, 69, 73, 77
intuition, 5, 19
    counter, 51–58
Israel, 46
iterative models, 32

Jefferson, Thomas, 64, 89
Johnson, Andrew, 4
Johnson, Lyndon B., 4
Jurassic Park (Crichton), 160–62, *160–62*

Kant, Immanuel, 20

karma, 11–12
Kennedy, John F., 4–7
Kepler, Johannes, 83
Klein bottle, 191–97, *191–93, 195–96*
    constructing, 191–93, *191–93*
    cube model of, 195–97, *195–96*
    fourth dimension and, 220–21, *221*
    holeless, 220–21
    one-sidedness of, 195, *195*
knots, 181–82, *181*
knotted rope, 215–19, *215–19*

Lakeside School, 47–48
Landon, Alfred, 43–44
Le Corbusier, 121, 129, 135, 139–40, *139*
Lennon, John, 133
Liber abaci (Fibonacci), 109
Lincoln, Abraham, 4–7
Literary Digest, 43–45
logarithmic spiral, 138, *138*
Lorentz, Edward N., 31–32, 38
Los Angeles, Calif., 83
lottery, 16
Lovelace, Augusta, 65
Lucas, Eduard, 117
Lucas numbers, 116–19, *117–18*
Lyell, Charles, 80

MacLaine, Shirley, 6
magnets, 36–37, *36–37*
mathematical chaos, 2, 23
mathematics, viii
    knots in, 181
    practicality and, 75
    and understanding the universe, 73–74
    see also numbers; topology
media bias, 45–46
mega, as prefix, 83
millions, 83–84
Möbius band, 184–89
    construction of, 185, 187–88
    cut in half, 189–90, *190*
    cutting open, 187
    identification diagram for, 188–89
    one-edgedness of, 186–87
    one-sidedness of, 186–87
    see also Klein bottle
Modernism, 128
Modular Man (Le Corbusier), 139–40, *139*
Mondrian, Piet, 121, 128–29
    destruction of volume concept of, 128–29
Monroe, Marilyn, 88
music, 130

national debt, 85, 233
National Enquirer, 7

natural numbers, 80–82
nature:
    chaos as fundamental feature of, 40
    synergy between mathematics and,
        100–119
nautilus shell, 138, *138*
New York, N.Y., 83
Nude Descending a Staircase (Duchamp),
    227–28, *228*
numbers, 63–77
    billions, 84–85
    counting, 79–80
    decimal, 261–63, *263*
    and degrees of separation, 88–89
    factoring, 67–69
    Goldbach Conjecture and, 75–76
    millions, 83–84
    natural, 80–82
    in playing card-table edge example, 90–99
    prime, *see* prime numbers
    quadrillions, 86–87
    real, 262
    repeated doubling of, 86–87
    size and, 64
    for their own sake, 74–75
    thousands, 83
    in *3x + 1 procedure*, 76–77
    trillions, 85–86
    unknowable, 232–33
    see also cryptography

one-dimensional line world, 204, *204*
    as continuum of points, 206–7, *206*
ones over ones fractions, 113–16, *113–15*
one-to-one pairings, 244, 245, 247–48, 257,
    260
    cardinality concept and, 239
    between decimal and counting numbers,
        261–63, *263*
    pairable means equal concept and, 233–34,
        234, 235
order-of-magnitude counting, 79–80
origami, 122
    see also paper-folding
Oswald, Lee Harvey, 4

paper-folding, 148–49, *148*
    doubling and, 86–87
    in *Jurassic Park* example, 160–62, *160–62*
    patterns in, 150–52, *151–52*
    and predicting next fractal image, 159–60,
        159–60
    regularity in sequences of, 153–55, *153–54*
    in Turing Machine program, 158
    valley and ridge sequences in, 149–51,
        149–50

Parade, La (Seurat), 128, *129*
parallel planes, 207–8, *207–8*
Parthenon, 123, 125–26, *126*, 128
patterns, 101–7, 108
　drip, 34–36
　in interlocking spirals, 104–6, *104–6*
　in paper-folding sequences, 150–52,
　　150–52
　as parallel spirals, 102–4
　in pineapple, 101–4, *102–3*, 107
　in pine cone, 106
　regularity of, 153–55, *153–54*
pendulums, 33–34, *33–35*
personality, genetics and, *see* twin studies
phi (φ), 115, 125, 136–37
pineapple, 107
　patterns in, 101–4, *102–3*, 107
pine cones, 106
Ping-Pong balls, 231, 241–44, *241*
Pinwheel Tiling, 143–45, *144*
planes:
　parallel, 207–8, *207–8*
　see also two-dimensional plane world
polling, 43–45
Pope, Alexander, 146
predictions; predicting:
　chaos and, 23, 32, 41
　coincidence and, 12–15
　of future, 23, 32, 41
　of next fractal image, 159–60, *159–60*
　Lorentz's discovery and, 31–32
　of U.S. elections, 43–45
　of weather, 21–22
Prelude to the Afternoon of a Faun (Debussy),
　123
Presley, Elvis, 88
prime numbers, 71–72
　as building blocks, 74
　public key encryption and, 71–72, 75
　twin, 76
Proclus, 78
proportions, 124–25, 127–28, *127–28*, 135,
　135
public key encryption, 69–73, 77
　prime numbers and, 71–72, 75
Pythagorean Theorem, 134–35

quadratic formula, 115
quadrillions, 86–87
quantum physics, 40
quaver knits, 130
quincunx, 50–51

rabbit-generation question, 109–10, *110*
randomness:
　in bouncing ball example, 50–51, *50*

coincidence and, 10–11
　in SAT responses, 48–49
　statistics and, 44–45, 48–51, *50*, *51*
ratio, *see* Golden Ratio
real numbers, 262
rectangle, *see* Golden Rectangle
recurrence sequence, 116–17, *117*, 139–40
Roosevelt, Franklin D., 43–44, 88
rubber undies trick, 171–73, *172*

Saint Jerome (da Vinci), 127–28, *127*
Saint Louis Cardinals, 236–37, 239
San Francisco Giants, 239
SAT, 48–51
Schwarzenegger, Arnold, 84
Security Dynamics, 73
Seurat, Georges, 128, *129*
Shakespeare, William, 10, 231
space:
　stacked three-dimensional, 207–9, *208*
　zero-dimensional, 204, 222
spinning pennies example, 52–53
spiral patterns, 102–4
　computer simulations of, 108
　Fibonacci numbers in, 108–9
　Golden Ratio and, 138, *138*
　interlocking, 104–6, *104–6*
　logarithmic, 138, *138*
　parallel, 102–4
statistics, 1, 42–61
　in air safety example, 56–58
　average income example and, 47–48
　balanced thinking and, 52
　bell-shaped curve and, 49–51, *50–51*
　biased sources and, 44–45
　counterintuition and, 51–58
　expected equality and, 54–55
　Gallup poll and, 44–45
　in HIV testing example, 58–60
　Literary Digest fiasco and, 43–45
　normal distribution and, 48–49
　personal bias and, 45–47
　randomness and, 44–45, 48–51, *50*, *51*
　in reunion scenario, 53–55
　risks and, 46–47
　SAT guessing and, 48–49
　in spinning pennies example, 52–53
　unintentional consequences and, 57–58, 60
　in U.S. presidential elections, 43–44
stock market, 12–15, 22
supercoiling, 183–84, *183–84*
symmetry, 143–45, *143–44*

tavern puzzles, 169–71, *170–71*
terrorism, 46
Thatcher, Margaret, 88

thousands, 83
three-dimensional world, 202–4, *203–4*, 205, *206*, 207
   as collection of parallel planes, 207–8, *207–8*
   perception, 224–25
3x + 1 procedure, 76–77
tiling, 142–45
time, 228–29
Tonight Show Starring Johnny Carson, 9
topology, 169–99
   cube model and, 195–97, *195–96*
   DNA and, 181–84, *183–84*
   doughnut-shape tasks and, 174–77, *175–77*
   Möbius band and, *see* Möbius band
   ring-disk problem and, 177–80, *177–80*
   rubber undies trick and, 171–73, *172*
   tavern puzzles and, 169–71, *170–71*
   trouser-inversion trick and, 166, 173, *173*
   unknots and, 181–82, *181*
translational symmetry, 143–45, *143–44*
triangle, *see* Golden Triangle
trillions, 85–86
trouser-inversion trick, 166, 173, *173*
Turing, Alan, 155
twin prime conjecture, 76
twins studies, 7–9

two-dimensional plane world, 204, 205–7, *205*, *209*
   boundaries of, 209–11, *210–11*
   as collection of lines, 207
   disappearing rabbit example of, 212–14, *212–14*
   knotted rope example and, 215–19, *215–19*

uncertainty, *see* chaos; coincidence
United States, 85
Universal Turing Machine, 155, 157, 163
universe, 64
   cube model of, 195–97, *195–96*
   elasticized, *see* topology
   mathematics and, 73–74
unknots, 181–82
Ussher, James, 79–80

volume, destruction of, 128–29

Weber, Max, 227, 229
"What if?" question, 199
"What next?" question, 199–200
World War II, 155

zero-dimensional space, 204, 222